Lecture Notes in Computer Science 2771

Edited by G. Goos, J. Hartmanis, and J. van Leeuwen

Springer
Berlin
Heidelberg
New York
Hong Kong
London
Milan
Paris
Tokyo

Uffe Kock Wiil (Ed.)

Computer Music Modeling and Retrieval

International Symposium, CMMR 2003
Montpellier, France, May 26-27, 2003
Revised Papers

Volume Editor

Uffe Kock Wiil
Aalborg University Esbjerg, Department of Software and Media Technology
Niels Bohrs Vej 8, 6700 Esbjerg, Denmark
E-mail: ukwiil@cs.aue.auc.dk

Cataloging-in-Publication Data applied for

A catalog record for this book is available from the Library of Congress.

Bibliographic information published by Die Deutsche Bibliothek
Die Deutsche Bibliothek lists this publication in the Deutsche Nationalbibliografie; detailed bibliographic data is available in the Internet at <http://dnb.ddb.de>.

CR Subject Classification (1998): H.3, H.4, H.5, H.2, I.2, C.3
Additional material to this book can be downloaded from http://extras.springer.com
ISSN 0302-9743
ISBN 3-540-20922-0 Springer-Verlag Berlin Heidelberg New York

Springer-Verlag is a part of Springer Science+Business Media

springeronline.com

Typesetting: Camera-ready by author, data conversion by Olgun Computergrafik
Printed on acid-free paper SPIN: 10931363 06/3142 5 4 3 2 1 0

Preface

This volume contains the final proceedings for the Computer Music Modeling and Retrieval Symposium (CMMR 2003). This event was held during 26–27 May 2003 on the campus of CNRS/Université de Montpellier II, located in Montpellier, France. CMMR is a new annual event focusing on important aspects of computer music. CMMR 2003 is the first event in this new series. CMMR 2003 was jointly organized by Aalborg University, Esbjerg in Denmark and LIRMM in France.

The use of computers in music is well established. CMMR 2003 provided a unique opportunity to meet and interact with peers concerned with the cross-influence of the technological and creative in computer music. The field of computer music is interdisciplinary by nature and closely related to a number of computer science and engineering areas such as information retrieval, programming, human computer interaction, digital libraries, hypermedia, artificial intelligence, acoustics, signal processing, etc. The event gathered several interesting people (researchers, educators, composers, performers, and others). There were many high-quality keynote and paper presentations that fostered inspiring discussions. I hope that you find the work presented in these proceedings as interesting and exciting as I have.

First of all, I would like to thank Marc Nanard, Jocelyne Nanard, and Violaine Prince for the very fruitful cooperation that led to the organization of this first event in the CMMR series. I would also like to thank my colleague Kirstin Lyon for her help in compiling these proceedings. Finally, this volume would not have been possible without the help of Springer-Verlag, Heidelberg. In particular, I would like to thank the editorial assistant, Anna Kramer, and the executive editor of the LNCS series, Alfred Hofmann.

November 2003

Uffe Kock Wiil
CMMR 2003 Program Chair

Organization

CMMR 2003 was jointly organized by LIRMM, CNRS/Université de Montpellier II, France and Aalborg University, Esbjerg, Denmark.

Organizing Committee

Chairs
Marc Nanard (LIRMM, France)
Jocelyne Nanard (LIRMM, France)

Program Committee

Chairs
Uffe Kock Wiil (Aalborg University, Esbjerg, Denmark)
Violaine Prince (LIRMM, France)

Members
Jens Arnspang (Aalborg University, Esbjerg, Denmark)
Antonio Camurri (University of Genoa, Italy)
Barry Eaglestone (University of Sheffield, UK)
Franca Garzotto (Politecnico di Milano, Italy)
Goffredo Haus (University of Milan, Italy)
David L. Hicks (Aalborg University, Esbjerg, Denmark)
Henkjan Honing (University of Nijmegen/University of Amsterdam, The Netherlands)
Kristoffer Jensen (University of Copenhagen, Denmark)
Andruid Kerne (Texas A&M University, USA)
Richard Kronland-Martinet (CNRS-LMA, Marseille, France)
Henrik Legind Larsen (Aalborg University, Esbjerg, Denmark)
Brian Mayoh (University of Aarhus, Denmark)
Peter J. Nürnberg (Technical University of Graz, Austria)
François Pachet (Sony Research Lab, France)
Esben Skovenborg (University of Aarhus, Denmark)
Bernard Stiegler (IRCAM, Paris, France)
Leonello Tarabella (CNR, Pisa, Italy)
Daniel Teruggi (INA, Paris, France)
Hugues Vinet (IRCAM, Paris, France)
Gerhard Widmer (University of Vienna, Austria)
Sølvi Ystad (CNRS-LMA, Marseille, France)

Sponsoring Institutions

Université de Montpelleier II, France
Région Languedoc Roussillon, France
CNRS, France
Aalborg University, Esbjerg, Denmark

Table of Contents

Some Projects and Reflections on Algorithmic Music

Rubén Hinojosa Chapel

Music Technology Group, Universitat Pompeu Fabra
Ocata 1, 08003 Barcelona, Spain
+34 93 542 2104
ruben.hinojosa@iua.upf.es

Abstract. Algorithmic Music Composition with computers (in real and non-real time) has had many approaches. It can be viewed from different points of view: scientific, technological, artistic or philosophical. In this paper the author introduces three projects developed between 1990 and 2000 in Havana, during his time as a professor, researcher and composer at the University of Arts of Cuba. He also exposes some philosophical reflections on algorithmic music systems, derived from his work at Havana and his current research in the Music Technology Group of the Universitat Pompeu Fabra.

Keywords: Algorithmic Composition, Real-Time Composition, Interactive Music Systems, Virtual Instruments, Computer Aided Composition, Algorithmic Music.

1 Introduction

In 1989 Cuban composer Carlos Fariñas (1934-2002) founded, with some colleagues, the *Estudio de Música Electroacústica por Computadora* (Studio for Electroacoustic and Computer Music) of the *Facultad de Música*, at the *Instituto Superior de Arte* (University of Arts) in Havana. One year later I started collaborating with the Studio as an undergraduate student of Computer Science.

From the beginning I was assigned a project on Algorithmic Music. Our knowledge on the field was almost null, so we started from scratch, provided only with a couple of articles, some historical and anecdotal references (Hiller, Xenakis...) and a great enthusiasm in doing our best. For one decade I have been working on Algorithmic Music composition and software development and experimentation. Now I have some experiences and reflections that would like to share with other people interested in the field.

2 Musical Fractals (1990-1994)

Our first idea was to create a system for generating musical structures automatically, a system without a direct connection with any known musical style. Nevertheless, as we will see later, we implemented melodic transformations from traditional counterpoint, and even new ones.

Fractal images were very popular at that moment; the musical experience based on fractals carried out by American composer Charles Dodge [1] was a starting point for

U.K. Wiil (Ed.): CMMR 2003, LNCS 2771, pp. 1–12, 2004.

our research. Dodge suggested a musical structure based on a metaphoric interpretation of the self-similarity concept. He departed from the Koch curve for building parallel voices that contain proportional relationships between them, similar to those existing between the triangles of the Koch curve.

We elaborated and implemented an algorithm based on the Dodge's interpretation of self-similarity. Our first system, *Musical Fractals*, needs as a seed data, a melody, a list of melody transformations, and some numerical values such as: number of voices, values that will affect the relationships between them, etc. It computes the "piece" in non-real time, generating up to four parallel voices. One of the voices is the original melody and its variations. We implemented some interesting features that proved, through practical experiences, its strength and weaknesses. Some of these features are [2]:

1. **Scales.** The program is able to use up to fifteen different musical scales, even a user defined one, for computing the whole "piece".
2. **Traditional melodic variations.** Melodic transformations from classical counterpoint are algorithmic procedures used along centuries of musical tradition. They are powerful tools for developing a melody, so we decided to test their potential inside a computer program.
3. **Non-traditional melodic variations.** Some non-traditional melodic variations were implemented, following very personal approaches. They are:
   - **Addition:** Randomly adds some notes to the melody without affecting the total length.
   - **Subtraction:** Randomly deletes some notes from the melody without affecting the total length.
   - **Reverse time:** It is like traditional "reverse", but it only reverses the note durations of the melody.
   - **Reverse pitch:** It is like traditional "reverse", but it only reverses the note pitches of the melody.
   - **Generation:** Uses a 1/f fractal noise generator for changing each pitch of the melody.
   - **Simulation:** Uses a particular approach, based on *Markov Chains*, for changing each pitch of the melody. The resulting melody sounds a little bit like the original one.
   - **Arpeggio:** Changes the notes whose duration is greater than or equal to a quarter-note, by an arpeggio of four notes, without affecting the total length of the melody. The algorithm uses interval values provided by the user.
   - **Logarithmic:** Changes every pitch by a new one, computed with a personal algorithm that uses the logarithmic function, and involves some existing pitches.

An interesting feature that proved good results is the possibility of applying not only isolated melodic variations to the melody, but a set of joined variations that conform a much complex transformation. For instance, think about applying the following variations in order: augmentation, arpeggio, simulation and diminution. The resulting melody is only one, not four. Of course, it is possible to obtain four different melodies too! A simple command interpreter was implemented for entering coding of complex transformations. Composers found this possibility as a new and powerful compositional tool [2].

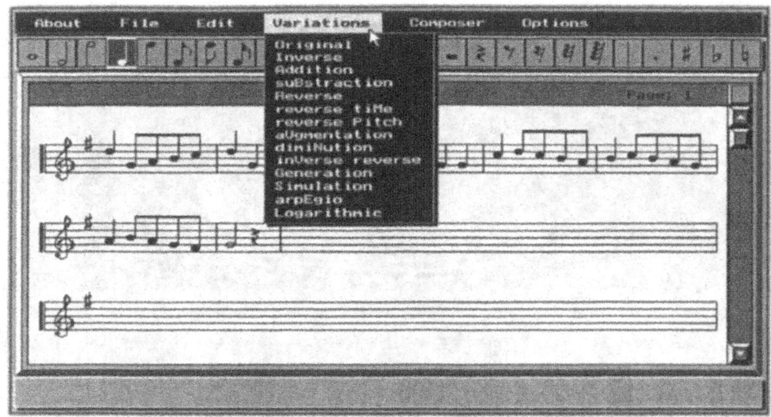

Fig. 1. A screenshot of Musical Fractals

A particular effort was the creation of an embedded score editor (by Claudio Daniel Ash) for entering the melody and for another upcoming projects. The resulting "piece", as well as the original melody and every transformation can be heard through a Roland MPU401 MIDI interface and an external sound module. Additionally, they can be saved to a Jim Miller's *Personal Composer* MIDI file.

Musical Fractals runs under MS-DOS and never was ported to MS Windows. It was awarded with some prizes. This project provided us with our first experiences in experimenting computer algorithms for music composition. One electroacoustic music piece was composed by Carlos Fariñas *(Cuarzo: Variaciones Fractales)*, which was premiered in 1991 in Havana, in the frame of the Festival of Contemporary Music [2], and has been played abroad. The author made also, in 1994, a short electronic piece entitled *ET llamando a casa*, which demonstrates some of the features of the computer music system.

2.1 Orbis Musicae (1993-1996)

In 1993 we started another project with new goals in mind. We were faced with the problem of creating an interactive music system for real-time performances. The idea came about through several ways, but a very influencing one was our personal meeting with Dr. Max Mathews in 1991, during the International Festival of Electroacoustic Music held in Varadero beach. There I had the opportunity to talk with him. Among several questions, I posed this one: How would you use a music made algorithmically? Mr. Mathews kindly answered:

"I would be interested in keeping an interaction with the algorithm; a part from the computer and a part from myself. I am interested in algorithms for improvising. With these algorithms, the musician and the computer play the music together. The algorithm chooses the notes, but the musician can select, among the options given by the program, the one he likes [3]."

With these ideas in mind we gave birth to *Orbis Musicae*, which acts like an instrument, where the musician controls different and variable parameters in real-time while the music is computed. The algorithm is quite straightforward:

Fig. 2. A screenshot of Orbis Musicae

There are twelve planets around the Sun moving each one on an elliptic trajectory. At the beginning each planet is assigned a grade from the chromatic scale. Then one or more triangles are placed over the orbital plane. When the planetary system starts moving, one or more planets visit the area of the triangles. As soon as a planet goes in or out from a triangle, a Note On or Note Off message is triggered, sounding on or off a MIDI controlled external sound source. The next time this planet goes inside the triangle, the original pitch assigned to it could remain the same, or could be changed to a new one, according to the initial choice of the musician. Each triangle has assigned a particular MIDI channel, so different MIDI programs (instruments) or sound modules can be controlled at the same time [4].

The musician can change the position and speed of the planets during the real-time performance. The configuration of the system, say: planet positions, planet speeds, assigned pitches, triangles and its assigned MIDI channels, can be saved internally at any time, and can be restored also whenever the user wants. *Orbis Musicae* uses ten memory banks, and follows the "total recall digital mixers" philosophy. The composer can use this feature for creating a scheme of configurations, which would be useful for planning the development of his piece in a sort of a score.

Orbis Musicae runs under MS-DOS and, as well as its predecessor, never was ported to MS Windows. An electroacoustic music piece was composed by Carlos Fariñas, who used its real-time capabilities for recording fragments of music in a sequencer. Later he took these fragments for creating a tape composition *(Orbitas Elípticas)*. This music work was played in 1994 in the Bourges's International Festival of Electroacoustic Music. Another electroacoustic music work was created by Cuban composer Roberto Valera, who used the software for his real-time piece *Hic et Nunc*, performed for the first time in 1996, with the assistance of myself, in the frame of the Havana's Festival of Contemporary Music.

Orbis Musicae has two essential properties:

1. It is a self-regulated system that has a personal behaviour. It can play itself endlessly without any human intervention.

2. The task of the human player is to influence the behaviour of the system, as if he were an instrument player. In fact, he is an instrument player. A player of a new kind of instrument, an active instrument. Traditional instruments always play a passive role, they react to the human gesture, but they are unable to offer the musician any musical idea by itself. At that time we used to name this kind of software active instrument, "virtual instrument".

In the middle of our investigation, we found previous experiences from other researchers whose works connect deeply to, and reinforce, the ideas we were working on. These experiences come, in one hand, from Louis and Bebe Barron, and on the other hand, from John Bischoff and Tim Perkis.

2.2 Louis and Bebe Barron

We found very interesting and pioneering the works done by Louis and Bebe Barron in New York, during the fifties of the past century. They intended to build new sonic models using the spontaneous electric evolution of some electronic circuits coupled between themselves, whose oscillation frequencies were placed in the audible range.

The main idea was to build series of active circuits with specific frequencies and transitory regime. By coupling these circuits to each other and influencing the behaviour of its neighbour circuits, it is possible to make changes to its own parameters. According to a partially predictable process, the union of synchronizing influences coming from its neighbour oscillators will modify the state of the oscillations of each circuit, so that they modulate their oscillations between themselves [5].

The first circuit state is dictated by external conditions, which can be changed at will. Leaving it to itself, the system of circuits follows an evolutionary process, which can be defined as the behaviour in reaction to external stimuli. This acoustic behaviour is modified according to the relationship and order established between the circuits, and confers personal characteristics to a particular considered system [5].

If we choose and study conveniently the parameters of those circuits, it could be possible to obtain an interesting sonic result, which could lead to the creation of an electronic music composition. Under this perspective [5], Louis and Bebe Barron made music for the cinema, especially for the films *Bells for the Atlantis, Electronic Jazz* and the science fiction film *Forbidden Planet* (1956). The soundtrack of this film is a wonderful example of artistic and avant-garde creation, and a remarkable example of the musical use of sound synthesis by modulation.

In the works by Louis and Bebe Barron we have found the notions of a self-regulated sound generation system, which owns a personal and autonomous sonic behaviour that can be controlled and changed by external influences in an interactive way. We have found also the same principle in the works by John Bischoff and Tim Perkis.

2.3 John Bischoff and Tim Perkis

On the CD *Artificial Horizon*, recorded between 1989 and 1990 by American composers John Bischoff and Tim Perkis, is exposed a sample of what they call "Music for New Software Instruments". In the CD booklet they express the philosophy of their music in the following terms:

"For us, composing a piece of music is building a new instrument, an instrument whose behavior makes up the performance. We act at once as performer, composer and instrument builder, in some ways working more like sculptors than traditional musicians. (...) There is another feature of the computer that attracts us: its ability to build systems of interaction with complex dynamics, systems only partially predictable, which can develop a unique "body" of their own. These woolly computer instruments can also be designed to respond to players' actions in new ways, creating a music which contains the trace of human gesture, in addition to having a degree of autonomy. In fact, for us, the distinction between composing a new piece of music and building a new instrument is not clear-cut: composing a piece of music for us IS building a new instrument, an instrument whose behavior makes up the performance. We act at once as performer, composer and instrument builder, in some ways working more like sculptors than traditional musicians. And in each case, the focus is on creating a system as open and alive as possible, bearing the precious marks of an individual character." [6]

And specifically talking about his 1978-80 piece *Audio Wave*, John Bischoff says: "AUDIO WAVE was written for pianist Rae Imamura (...). My idea was to make a live computer piece for Rae where both of her hands would be continually active, as in her conventional keyboard playing, but where her actions would serve to influence an ongoing musical output rather than have the task of initiating each sound." [7]

When we knew about the principles behind the works by the Barrons, Bischoff and Perkis, and after looking back to our experiences, we felt that we had found what path to travel through. So, we decided to build an interactive algorithmic music system for real-time performances, not-based on any known musical style, which could act as an active instrument (self-regulated system) "where the user's actions would serve to influence an ongoing musical output rather than have the task of initiating each sound". We were looking for a more flexible Virtual Active Instrument. Then the *Fractal Composer* project was started at the EMEC/ISA.

3 Fractal Composer (1996-2000)

Fractal Composer [8] is intended to be a virtual musical instrument for real-time performances. It plots chaotic attractors, dynamic systems and some related formulas, and makes music from these calculations while the musician introduces changes to musical parameters and listens to the results, all of this in real-time.

The system, which runs under MS Windows, features seven different fractal formulas and related algorithms for tone generation, which combine six ways of mapping pixel colors into pitches, and four note-duration or rhythms. Up to four interdependent voices may be used, conducted through three different manners or styles. Each voice owns its loudness or dynamics, its pitch limits (range) and scale. This program offers twenty-four different scales, including nine user defined ones.

The user may have control over some MIDI functions like: program changes, modulation and panning. Each voice can be moved from left to right or vice versa, automatically, at the speed the user chooses. Although program names are showed in the General MIDI convention, of course it is possible to use any non-GM external sound module. In addition to this, it is possible to load a digital sound file and play it together with the MIDI fractal music.

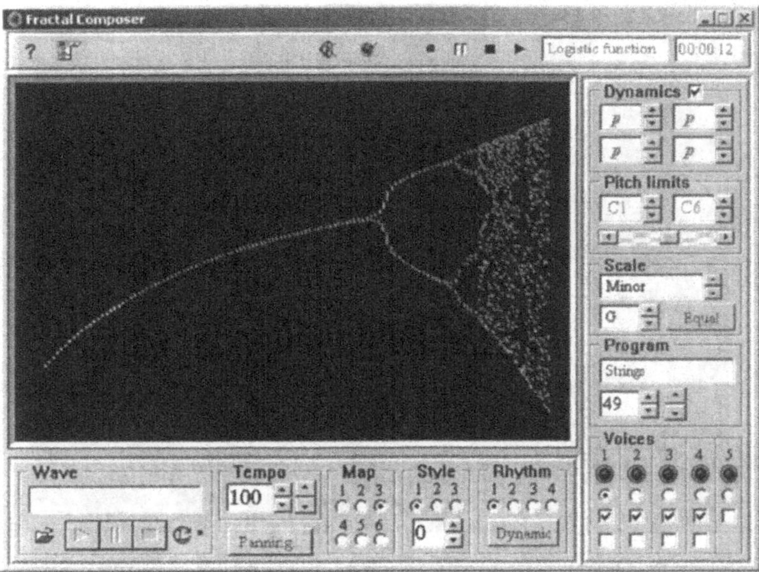

Fig. 3. A screenshot of Fractal Composer

While *Fractal Composer* creates music in real-time, it is possible to save the resulting music, with all the changes (performance) the user has done, in a Standard MIDI File. This lets him edit his music in any sequencer or music notation software that support SMF.

The musician can store all the settings in a configuration file to be recalled later, in another session. This means that the player doesn't lose his fractal type nor its parameters, voices selected, patches, dynamics, scales, and even his own scales. A chronometer appears in the upper right-hand corner of the display to inform the performer about the duration of his piece as time goes by.

As Xenakis said in 1971 in his book *Formalized Music*: "With the aid of electronic computers, the composer becomes a sort of pilot: pressing buttons, introducing coordinates, and supervising the controls of a cosmic vessel sailing in the space of sound, across sonic constellations and galaxies that could formerly be glimpsed only in a distant dream." [9]

The author wrote three electroacoustic music pieces with this system:

1. *El fin del caos llega quietamente*, which is intended to demonstrate that mathematics can also be a path to music. It was created entirely in real-time from the calculus of the *Logistic Function*, and recorded in one pass with no overdubbing.
2. *Satélites*. Basic sonic material was created in real-time from the calculus of the Henon Map. It was premiered in the XII Havana's Festival of Contemporary Music, in 1997.
3. *Oro-iña*, a real-time performance for computer, Afrocuban drum set and two dancers. It was played for the first time in 1998, in the frame of the International Festival of Electroacoustic Music held in Havana.

Fig. 4. Picture of the Oro-Iña performance (Photo: Archie)

4 Some Theoretical Reflections

4.1 In Search of a Satisfactory Algorithm

Algorithmic Composition researchers have tried different approaches for handling the music elements and for generating musical structures. Traditionally one of the most important elements of music has been melody. Many algorithms and models for "composing" melodies have been developed, from 1/f fractal noise to rule based or constraint programming.

Cuban composer Carlos Fariñas (1934-2002) used to say that every melody but those from monodic systems, always has an implicitly harmonic context. So, according to this idea, every random procedure for creating melodies should take this principle into account. It has no sense to look for an algorithm for creating "beautiful" or "inspired" melodies without influencing the random process by a harmonic progression.

On the other hand, when researchers intend to mimic a known music style, it is often known what elements, characteristics or procedures they should model, but what path should be follow in order to generate satisfactory musical structures not-based on any known musical style? Musicology and music composition tradition have the answer. When we were looking for a solution, composer Fernando Rodríguez (Archie) came up to us and replied: "you should try to model Analogy and Contrast".

"The notions of foreground and background (...) are critical in controlling musical flow. If similarity is in the foreground, the listener will perceive the music as continuing uninterrupted; if difference is more prominent, then the perception will be one of contrast. (...) When contrast is in the foreground, it is introduced to avoid boredom, and to deepen the listener's experience. Contrast creates emotional breadth, setting off ideas and heightening relief and definition of character. (...) Musically, when we hear familiar material in new contexts, its meaning is enriched." [10]

These reflections around melody and structure only refer to our western music tradition. It could be possible that they do not match with music traditions from other different cultures.

4.2 In Search of a Definitive Composition System

Throughout the music history, composers have developed techniques, most of them algorithmic procedures, to handle all the elements of music, say: melody, harmony, rhythm, timbre, articulation, form... What a composition system does is to apply these techniques, and even new or personal ones, to the material provided by the user, i.e., the composer. The ways to handle the elements of music are so many, almost infinite so, from our point of view, it is impossible to find a definitive computer composition system [11].

Every music algorithm leaves its fingerprinting in the sonic result of its execution. It has no sense to look for an universal algorithm for composing any known music style. Composers use many algorithms or algorithmic procedures everyday, and the doors for creating new compositional procedures and new music styles are always opened, though it's no easy to travel it through. Music composition involves creativity, which is impossible to lock in a scientific model. It always flies away beyond our imagination.

4.3 Two Reflections About Authorship

In [12] I found an interesting question that made me think: "(...) if an algorithm faithfully represents an artist's creative process, what is the difference between music produced by the artist and music produced by the algorithm?"

Algorithmic composition leads to the following situation: the user gives instructions to a computer to conceive an object (music). After a while, the man receives this object from the machine. So, what now? He says: "this is my own work". Has the man stolen the object from the machine? Does this object belong to the computer?

Do not forget who has mentally conceived that object before its physical existence. Man has thought about that object, with more or less precision, before giving instructions to the machine. So, the computer has the task to give birth to the object dreamed by the man. When an artist designs a monumental sculpture, it is built by several (or even many) workers, but nobody has doubts about the authorship of the sculpture. Who is the author of the *Sagrada Familia* temple? Who denies it is Antonio Gaudí? Who denies the authorship of the *Tour Eiffel* to Gustave Eiffel?

Computers only simulate, through very strict instructions from the man, some elements of the human thinking. During the creative process, they can contribute some things to the task commanded by the man, but they only can contribute things that were thought before, things that were mentally conceived previously by the man. They cannot contribute things unconceived by the man, because they have no will nor awareness. That is the difference between music produced by the artist and music produced by the algorithm.

The man conceives and programs creation strategies, which imitate his possibilities, skills and knowledge. So, his personality will be present in the machine's results. Computers have no special artistic skills or virtuosity. They only have a representation of the skills and knowledge from the man. They are only able to mimic those human properties.

Can a machine express its individuality, its own personality, its own subjectivity? These qualities are not properties of a computer, so they cannot be expressed. Only the man can express his individuality, personality and subjectivity, from the moment he selects and gives instruction to the machine, from the very instant he conceives a music program, or when he configures the options of the software. Machines impregnate with some logic and formal characteristics the result of its computations, but the man is who gives imagination to those calculi, the man is who transmits his human sensibility with the help of a computer, and he is who transforms in art the science that could exists in automatic creations.

I have found also in [12] the following interesting statement: "(...) music produced by algorithmic composition is considered somehow inferior not because it was produced by an algorithm, but because it is someone else's music–it belongs to the *designer* of the algorithm, and not to the user of the algorithm."

If we accept this statement as a valid one, maybe it should be said: *Wozzek* does not belong to Alban Berg (the *user* of the twelve-tone algorithmic procedures) but to Arnold Schoenberg (the *designer*). Traditional non-computer algorithmic methods are really compositional procedures, which are always adapted by the composer to his mental scheme, to his personal point of view about music, and to his own experience and skills.

When the user configures the options of any algorithmic composition system, and gives it the seed data, he transmits his own personality, as well as when he uses any conventional algorithmic procedure, or even a rule-based music composition formalism like traditional counterpoint. So, we firmly believe that music composed with the aid of an algorithmic composition system, belongs to the user.

4.4 Algorithmic Music Composition: Why?

Due to the wide range of possibilities offered by computers and other electronic music devices, which are sometimes exaggerated, it is often though erroneously that usual music knowledge is unnecessary for making music with those equipments. We think computers are a powerful tool for the musician. They will help the artist in developing his ideas, in stimulating his imagination, in speeding up some technical procedures of music composition. Computers enrich the compositional process, but they will not provide the user an unexisting talent. Nevertheless, they are able to stimulate the development of an undiscovered talent or innate musical capabilities [13].

Finally, I would like to point out some general ideas related with algorithmic music composition systems I have compiled:

1. These systems stimulate the composer's creative imagination in a very new and promising way, with lots of possibilities.
2. Composition programs can handle much more data and much faster than a human composer. They let him think in a high level of abstraction, leaving low-level details to the computer.
3. They are a door for searching new aesthetic concepts, new sonic conceptions and new ways of organizing sounds. So, they are a path for music development.
4. These systems allow scientific verification of music theories, when it is intended to simulate a known musical style in order to analyse and study it.
5. They allow to better know how musical processes take place in the human mind, so they let us know better the nature of the human being.

5 Present and Future Work

In 2001 I was granted a scholarship from the *Universitat Pompeu Fabra* in Barcelona, for making my doctor degree in Computer Science and Digital Communication. Now I have a good opportunity to learn new things, to work on new projects and to develop the ideas I have been working on in Cuba since 1990.

In December 2002 was created inside the MTG (*Music Technology Group*) led by Dr. Xavier Serra, the IST (*Interactive Systems Team*). Having Sergi Jordà as Project Manager, the IST is integrated by: Alvaro Barbosa, Gunter Geiger, Martin Kaltenbrunner, José Lozano and myself. Now we work on a new project named *reacTable*∗, which puts together most of the research interests and know-how of the IST members.

ReacTable∗ is a project that activates important interdisciplinary research in the field of Computer Music, which significantly departs from the MTG traditional work based on signal processing techniques. Some of the areas of research involved are algorithmic composition and real-time music creation / composition. For the near future we hope to integrate our experiences in the development of this new project.

6 Conclusions

Algorithmic music composition with computers (in real and non-real time) has had many approaches [9], [14–24]. It can be viewed from several points of view: scientific, technological, artistic or philosophical. We have introduced three projects developed between 1990 and 2000 at the EMEC/ISA in Havana. The first one generates musical structures in non-real time, while the other ones also generate musical structures but in real-time, in an interactive way. Neither is based on known musical styles, though they use basic musical concepts or technical procedures. To my mind, a concluding idea is the development of an interactive algorithmic music system for real-time performances, not-inspired (no mimic) on any known musical style, which could act as an active (self-regulated) instrument "where the user's actions would serve to influence an ongoing musical output rather than have the task of initiating each sound". Finally, we have exposed some theoretical reflections. I hope this paper be useful for the development of discussions and ideas related with the topics discussed here.

Acknowledgments

The author would like to thank the following people: composers Carlos Fariñas, Roberto Valera and Fernando Rodríguez (Archie), and all the students of Music Composition at the *Facultad de Música* of the *Instituto Superior de Arte* in Havana, for their contribution with ideas, suggestions, comments, discussions, critics and direct experiences with the projects I was working on during my time as researcher and professor in that place (1990-2000). Carlos A. Gonzalez Denis for having introduced me to fractal theory. Composers John Bischoff and Chris Brown for their bibliographic contribution. Composer Gabriel Brncic for his encourages and support for making my Doctor degree at the *Universitat Pompeu Fabra* in Barcelona. Sergi Jordà for his valuable ideas and experiences as the leader of our Interactive Systems Team and the *reacTable*∗ project. Dr. Xavier Serra for his support to my stay in the Music Technology Group and the Ph.D. program at the *Universitat Pompeu Fabra*, and for supporting the *reacTable* project.

References

1. Dodge, Ch., Bahn, C. R.: Musical Fractals: Mathematical Formulas Can Produce Musical as well as Graphic Fractals. Byte, June 1986, p. 185-196
2. Hinojosa Chapel, Rubén: Sistemas de Ayuda a la Composición Musical. Thesis work, Faculty of Mathematics and Computer Sciences, University of Havana (1993)
3. Hinojosa Chapel, Rubén: Acerca de los Instrumentos Electrónicos, la Música Electroacústica y las Computadoras. Banco de Ideas Z / Instituto Superior de Arte, Havana (1995)
4. Hinojosa Chapel, Rubén: Sistemas de Ayuda a la Composición Musical: la Experiencia del EMEC. Proceedings of the I International Congress of Informatics on Culture, International Convention INFORMATICA '94, Havana, February 1994
5. Moles, Abraham: Les Musiques Expérimentales. Editions du Cercle d'Art Contemporain (1960)
6. Bischoff, J., Perkis, T.: Artificial Horizon. CD booklet, Artifact Recordings (1990)
7. Bischoff, John: Software As Sculpture: Creating Music From the Ground Up. Leonardo Music Journal, Vol. 1 No. 1, 1991
8. Hinojosa Chapel, Rubén: Fractal Composer: un Instrumento del Siglo XXI. Proceedings of the IV International Congress of Informatics on Culture, International Convention INFORMATICA 2000, Havana, May 2000
9. Xenakis, Iannis: Formalized Music: Thought and Mathematics in Composition. Rev. ed. Stuyvesant, New York, Pendragon Press (1992)
10. Belkin, Alan: A Practical Guide to Musical Composition `http://www.musique.umontreal.ca/personnel/Belkin/bk/4a.html#2)%20Contrast`
11. Hinojosa Chapel, Rubén: Music Generation Panel (A critical review). Workshop on Current Research Directions in Computer Music, Barcelona, Nov 15-16-17, 2001, Institut Universitari de l'Audiovisual, Universitat Pompeu Fabra `http://www.iua.upf.es/mtg/mosart/panels/music-generation.pdf`
12. Jacob, Bruce L.: Algorithmic Composition as a Model of Creativity. Organised Sound, volume 1, number 3, December 1996 `http://www.ee.umd.edu/~blj/algorithmic\_composition/algorithmicmodel.html`
13. Hinojosa Chapel, Rubén: Entre Corcheas y Electrones. Por Esto!, Mérida, Yucatán, Mexico, July 31, 1996
14. Alpern, Adam: Techniques for Algorithmic Composition of Music `http://hamp.hampshire.edu/~adaF92/algocomp/algocomp95.html`
15. Brün, Herbert: From Musical Ideas to Computers and Back. In: Harry B. Lincoln (Ed.), The Computer and Music Ithaca, Cornell University Press (1970)
16. Cope, David: Computers and Musical Style. Madison, WI: A-R Editions (1991)
17. Hiller, L., L. Isaacson: Experimental Music. McGraw-Hill Book Company, Inc., New York (1959)
18. Koenig, G. M.: Project One. Electronic Music Report 2. Uthrecht: Institute of Sonology, 1970. Reprinted 1977, Amstermdam: Swets and Zeitlinger
19. Kunze, Tobias: Algorithmic Composition Bibliography `http://ccrma-www.stanford.edu/~tkunze/res/algobib.html`
20. Maurer IV, John A.: A Brief History of Algorithmic Composition `http://ccrma-www.stanford.edu/~blackrse/algorithm.html`
21. Papadopoulo, G., Wiggins, G.: AI Methods for Algorithmic Composition: A Survey, a Critical View and Future Prospects. `http://www.soi.city.ac.uk/~geraint/papers/AISB99b.pdf`
22. Roads, Curtis: The Computer Music Tutorial. The MIT Press, Second printing (1996)
23. Zicarelli, David.: M and Jam Factory. Computer Music Journal 11(4), 1987, p. 13
24. More references: `http://www.flexatone.com/algoNet/index.html`

Real-Time Beat Estimation Using Feature Extraction

Kristoffer Jensen and Tue Haste Andersen

Department of Computer Science, University of Copenhagen
Universitetsparken 1
DK-2100 Copenhagen, Denmark
{krist,haste}@diku.dk,
http://www.diku.dk/~krist

Abstract. This paper presents a novel method for the estimation of beat interval from audio files. As a first step, a feature extracted from the waveform is used to identify note onsets. The estimated note onsets are used as input to a beat induction algorithm, where the most probable beat interval is found. Several enhancements over existing beat estimation systems are proposed in this work, including methods for identifying the optimum audio feature and a novel weighting system in the beat induction algorithm. The resulting system works in real-time, and is shown to work well for a wide variety of contemporary and popular rhythmic music. Several real-time music control systems have been made using the presented beat estimation method.

1 Introduction

Beat estimation is the process of predicting the musical beat from a representation of music, symbolic or acoustic. The beat is assumed to represent what humans perceive as a binary regular pulse underlying the music. In western music the rhythm is divided into measures, e.g. pop music often has four beats per measure. The problem of automatically finding the rhythm include finding the time between beats (tempo), finding the time between measures, and finding the phase of beats and measures. This work develops a system to find the time between beats from a sampled waveform in real-time. The approach adopted here consists of identifying promising audio features, and subsequently evaluating the quality of the features using error measures.

The beat in music is often marked by transient sounds, e.g. note onsets of drums or other instruments. Some onset positions may correspond to the position of a beat, while other onsets fall off beat. By detecting the onsets in the acoustic signal, and using this as input to a beat induction model, it is possible to estimate the beat.

Goto and Muraoka [1] presented a beat tracking system, where two features were extracted from the audio based on the frequency band of the snare and bass drum. The features were matched against pre-stored drum patterns and resulted in a very robust system, but only applicable to a specific musical style.

U.K. Wiil (Ed.): CMMR 2003, LNCS 2771, pp. 13–22, 2004.

Later Goto and Muraoka [2] developed a system to perform beat tracking independent of drum sounds, based on detection of chord changes. This system was not dependent on the drum sounds, but again limited to simple rhythmic structures. Scheirer [3] took another approach, by using a non-linear operation of the estimated energy of six bandpass filters as feature extraction. The result was combined in a discrete frequency analysis to find the underlying beat. The system worked well for a number of rhythms but made errors that related to a lack of high-level understanding of the music. As opposed to the approaches described so far Dixon [4] built a non-causal system, where an amplitude-based feature was used as clustering of inter-onset intervals. By evaluating the inter-onset intervals, hypotheses are formed and one is selected as the beat interval. This system also gives successful results on simpler musical structures.

The first step of this work consists of selecting an optimal feature. There are a very large number of possible features to use in segmentation and beat estimation. Many audio features are found to be appropriate in rhythm detection systems, and one is found to perform significantly better. The second step involves the introduction of a high-level model for beat induction from the extracted audio feature. The beat induction is done using a running memory module, the beat probability vector, which has been inspired by the work of Desain [5].

The estimation of beat interval is a first step in the temporal music understanding. It can be used in extraction and processing of music or in control of music. The beat detection method presented here is in principle robust across music styles. One of the uses of the beat estimation is in beat matching, often performed by DJs using contemporary electronic and pop music. For this reason, these music styles has mainly been used in the evaluation. The system is implemented in the open source DJ software Mixxx [6] and has been demonstrated together with a baton tracking visual system for the use of live conducting of audio playback [7].

2 Audio Features

The basis of the beat estimation is an audio feature that responds to the transient note onsets. Many features have been introduced in research of audio segmentation and beat estimation. Most features used here have been recognized to be perceptually important in timbre research [8]. The features considered in this work are: amplitude, spectral centroid, high frequency energy, high frequency content, spectral irregularity, spectral flux and running entropy, all of which have been found in the literature, apart from the high frequency energy and the running entropy.

Other features, such as the vector-based bandpass filter envelopes [3], or mel-cepstrum coefficients have not been evaluated. Vector-based features need to be combined into one measure to perform optimally, which is a non-trivial task. This can be done using for instance artificial neural nets [9] that demands a large database for training, or by summation [3] when the vector set is homogeneous.

Most features indicate the onsets of notes. There is, however, still noise on many of the features, and the note onsets are not always present in all features. A method to evaluate and compare the features is presented in section 3, and used in the the selection of the optimal feature. In the following paragraphs, a number of features are reviewed and a peak detection algorithm is described.

2.1 Features

The features are all, except the running entropy, computed on a short time Fourier transform with a sliding Kaiser window. The magnitude $a_{n,k}$ of block n and FFT index k is used. All the features are calculated with a given block and step size (N_b and N_s respectively).

The audio features can be divided into absolute features that react to specific information weighted with the absolute level of the audio and relative features that only react to specific information. The relative features are more liable to give false detection in weak parts of the audio.

The amplitude has been found to be the only feature necessary in the tracking of piano music [10]. This feature is probably useful for percussive instruments, such as the piano or guitar. However, the amplitude feature is often very noisy for other instruments and for complex music.

Fundamental frequency is currently too difficult to use in complex music, since it is dependent on the estimation method. It has been used [9] in segmentation of monophonic audio with good results, though.

One of the most important timbre parameters is the spectral centroid (brightness) [11], defined as:

$$SC_n = \frac{\sum_{k=1}^{N_b/2} k a_{n,k}}{\sum_{k=1}^{N_b/2} a_{n,k}}. \tag{1}$$

The spectral centroid is a measure of the relative energy between the low and high frequencies. Therefore it seems appropriate in the detection of transients, which contain relatively much high frequency energy.

An absolute measure of the energy in the high frequencies (HFE) is defined as the sum of the spectral magnitude above 4kHz,

$$HFE_n = \Sigma_{k=f_{4k}}^{N_b/2} a_{n,k}. \tag{2}$$

where f_{4k} is the index corresponding to 4 kHz.

Another absolute measure, the high frequency content (HFC) [12] is calculated as the sum of the amplitudes and weighted by the frequency squared,

$$HFC_n = \Sigma_{k=1}^{N_b/2} k^2 a_{n,k}. \tag{3}$$

These features are interesting because they indicate both high energy, but also relatively much high frequency energy.

The spectral irregularity (SPI), calculated as the sum of differences of spectral magnitude in one block,

$$SPI_n = \Sigma_{k=2}^{N_b/2} \left| a_{n,k} - a_{n,k-1} \right|, \tag{4}$$

and the spectral flux (SPF), calculated as the sum of spectral magnitude differences between two adjoining blocks,

$$SPF_n = \Sigma_{k=1}^{N_b/2} \left| a_{n,k} - a_{n-1,k} \right|, \tag{5}$$

are two features known from the timbre perception research. These features give indication of the noise level and the transient behavior that are often indicators of beats.

Note onsets can be considered as new information in the audio file. Therefore the running entropy, calculated on a running histogram of the 2^{16} quantization steps is considered. First the probability of each sample value is estimated for one block,

$$H_n(s(l)) = H_n(s(l)) + \frac{1}{N_b}, l = (n-1)N_s + 1 \cdots (n-1)N_s + N_b, \tag{6}$$

then the probability is updated with $1 - W_h$,

$$H_n = W_h H_n + (1 - W_h) H_{n-1}, \tag{7}$$

and finally the entropy in bits is calculated,

$$Ent_n = -\Sigma_{k=1}^{2^{16}} H_n(k) \log_2(H_n(k)). \tag{8}$$

These are the features evaluated in this work. The note-onsets are considered to occur at the start of the attacks, but the features generally peak at the end of the attacks. To compensate for this delay the time derivative is taken on the features. The second derivative is taken on the running entropy. The maximum of the derivative of the amplitude has been shown to be important in the perception of the attack [13]. In addition, the negative values of each feature are set to zero.

An example of the resulting time-varying extracted features can be seen in fig. 1 for a contemporary music piece[1]. On the figure manually marked note onsets are indicated by dashed lines. It is clear that most features peak at the note onsets. There is, however, still noise on many of the features, and some of the note onsets are not always present in the features.

2.2 Peak Detection

The features considered in the previous section all exhibit local maximums at most of the perceptual note onsets. To identify a note onset from a given feature a peak detection algorithm is needed. The peak detection algorithm used here chooses all local maximums, potentially using a threshold,

$$p = (F_{n-1} < F_n > F_{n+1}) \wedge (F_n \geq th) \tag{9}$$

where F is an arbitrary audio feature. In addition to the peak detection, a corresponding weight, w_k is also calculated at each peak k (at the time t_k),

[1] Psychodelik. Appearing on LFO - Advance (Warp 039), January 1996.

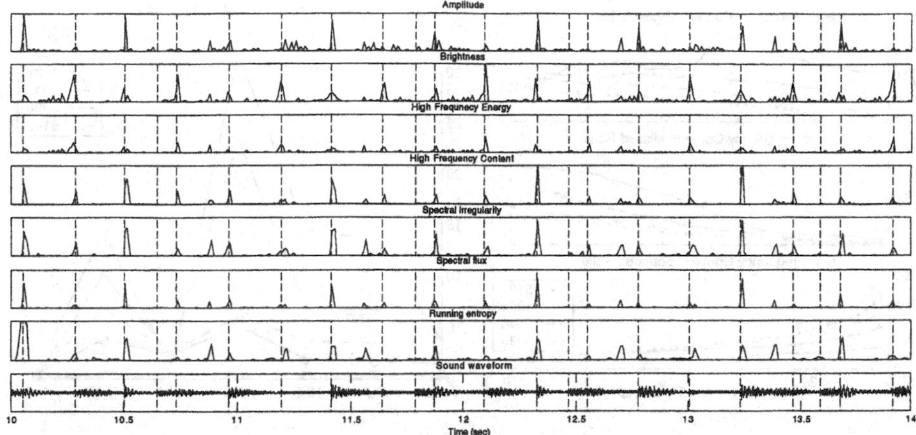

Fig. 1. Audio features from the LFO -Psychodelik piece (excerpt) as function of time. The features are shown at arbitrary scales. The vertical dashed lines indicate the manual marked transients.

corresponding to the time steps where p is true. This weight is later used in the beat probability vector, and in the detection of the phase of the beat. The threshold is used in the selection of the optimal feature, but not in the final beat estimation system.

3 Feature Analysis

To compare features, different musical pieces has been analyzed manually by placing marks at every perceptual note onset. The marking consists in identifying the note onsets that are perceptually important for the rhythmic structure. These note onsets are generally generated by the hi-hat and bass drum and any instrument with a transient attack. In practice, some parts of the pieces lack hi-hat and bass drum, and the rhythmic structure is given by other instruments. The manual marking of the note onsets in time has an error estimated to be below 10 msec. In all eight musical pieces were used, with an average of 1500 note onsets per piece.

These manual marks are used as basis for comparing the performance of the various features. In order to select the optimum feature, three different error measures are used, based on matched peaks, that is peaks located within a time threshold (20 msec) to a manual mark. An unmatched peak is located outside the time threshold from a manual mark.

3.1 Error Measures

To find the signal to noise the value of a matched (P), or unmatched ($\hat{P}$) peak is calculated as the sum of the feature at both sides of the peak where the slope

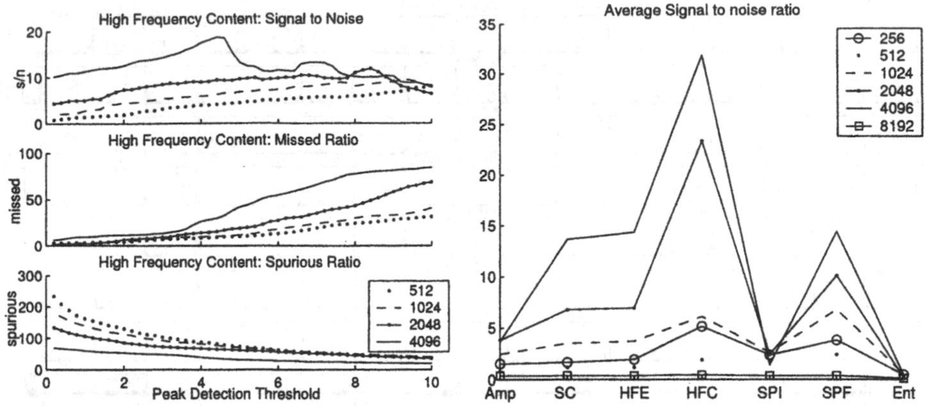

Fig. 2. Left: Example of error measures calculated using different block sizes of the HFC feature for the piece Train to Barcelona. Right: Average signal to noise for fixed threshold, and all music pieces for many block sizes and features.

is continually descending from the peak center. The signal to noise ratio is then calculated as,

$$s_n = \frac{\sum_{n=0}^{N_{matched}} P_n}{\sum_{n=0}^{N_{unmatched}} \hat{P}_n}. \tag{10}$$

The missed ratio is calculated as the number of manual marks minus the number of matched peaks, divided by the number of manual marks,

$$R_{missed} = \frac{N_{marked} - N_{matched}}{N_{marked}}, \tag{11}$$

and the spurious ratio is calculated as the number of unmatched peaks, divided by the number of manual marks,

$$R_{spurious} = \frac{N_{unmatched}}{N_{marked}}. \tag{12}$$

3.2 Analysis and Selection of Feature

In order to evaluate the features the error measures are now calculated on the music material using a varying peak detection threshold. An example of the error measures for the piece Train to Barcelona[2] is shown in the left part of fig. 2. For low thresholds, there are few missed beats, and for high peak detection threshold, there are many missed beats. The spurious beats (false indications) behave in the opposite way, for low thresholds there is up to several hundred percents, whereas the spurious ratio is low for high peak detection thresholds.

[2] By Akufen. Appearing on Various - Elektronische Musik - Interkontinental (Traum CD07), December 2001.

Under these conditions it is difficult to select an optimum peak detection threshold, since both low missed and spurious ratio is the optimization goal and they are mutually exclusive. The signal to noise ratio generally rises with the peak detection threshold, which indicates that the few found peaks contain most of the energy for the high thresholds. There seem to be no optimum way of selecting the threshold.

An analysis of the error values for all features and music pieces gives no clear indication of the best feature. Therefore a different approach has been used.

Initial tests have shown that the beat estimation method presented in the next section need at least 75% of the note onsets to perform well. The threshold for 75% matched beats (25% missed) is therefore found for each features/block size pair and music piece. The signal to noise ratio is then found for this threshold. The average signal to noise ratio is calculated for all music pieces. The result is shown in the right part of fig. 2.

Several results can be obtained from the figure. First, it is clear that the extreme block sizes, 256, 512, and 8192 all perform inadequately. Secondly, several features also perform poorly, in particular the amplitude, the spectral irregularity, and the entropy. The best features are the spectral centroid, the high frequency energy, the high frequency content and the spectral flux. The HFC performs significantly better than the other features, in particular for the block sizes 2048 and 4096, which has the best overall signal to noise ratio.

4 Beat Estimation

The analysis of the audio features has permitted the choice of feature and feature parameters. There is, however, still errors in the detected peaks of the chosen features. As described in other beat estimation systems found in the literature, a beat induction system, that is a method for cleaning up spurious beats and introducing missing beats, is needed. This could be, for instance, based on artificial neural nets, as in [9], but this method demands manual marking of a large database, potentially for each music style. Another alternative is the use of frequency analysis on the features, as in [3], but this system reacts poorly to tempo changes.

Some of the demands of a beat estimation system are stability and robustness. Stability to ensure that the estimation is yielding low errors for music exhibiting stationary beats and robustness to ensure that the estimation continues to give good results for music breaks without stationary beats. In addition, the system should be causal, and instantaneous. Causal to ensure real-time behavior, and instantaneous to ensure fast response.

These demands are fulfilled by the use of a memory-based beat probability vector that is based on the model of rhythm perception by Desain [5]. In addition a tempo range is needed to avoid the selection of beat intervals that do not occur in the music style. The tempo is chosen in this work to lie between 50 and 200 BPM, which is similar to the constraints used in [3].

4.1 Beat Probability Vector

The beat probability vector is a dynamic model of the beat intervals that permits the identification of the beat intervals from noisy features. The probability vector is a histogram of note onset intervals, as measured from the previous note onset. For each new note onset the probability vector $H(t)$ is updated (along with its neighboring positions) by a Gaussian shape at the intervals corresponding to the distance to the previous peak. To maintain a dynamic behavior, the probability vector is scaled down at each time step. At every found peak k the peak probability vector is updated,

$$H(t) = W^{t_k - t_{k-1}} H(t) + G(t_k - t_{k-1}, t), t = 0 \ldots \infty \qquad (13)$$

where W is the time weight that scale down the probability of the older intervals, and G is a Gaussian shape which is non-zero at a limited range centered around $t_k - t_{k-1}$. The current beat interval is identified as the index corresponding to the maximum in the beat probability vector, or, alternatively, to $t_k - t_{k-1}$ if the interval is located at the vicinity of the maximum in the beat probability vector.

The memory of the beat probability vector allows the detection of the beat interval in breaks with missing or alternative rhythmic structure. An instantaneous reaction to small tempo changes is obtained if the current beat interval is set to the distance between peaks at proximity to the maximum in the vector.

In [5] multiples of the intervals are also increased. Since the intervals are found from the audio file in this work, the erroneous intervals are generally not multiples of the beat. Another method must therefore be used to identify the important beat interval.

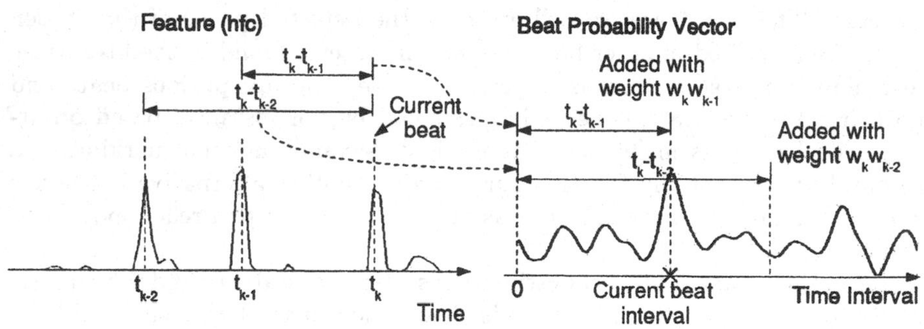

Fig. 3. *Selection of beats in the beat probability vector. For each new peak (left), a number of previous intervals are scaled and added to the vector (right). The maximum of the beat probability vector gives the current beat interval.*

4.2 Update with Multiple Intervals

To avoid a situation where spurious peaks create a maximum in the probability vector with an interval that does not match the current beat, the vector is updated in a novel way. By weighting each new note and taking multiple previous note onsets into account, the probability vector $H(t)$ is updated with N previous weighted intervals that lie within the allowed beat interval,

$$H(t) = H(t) + \Sigma_{i=1}^{N} w_k w_{k-i} G(t_k - t_{k-i}, t), t = 0 \ldots \infty \qquad (14)$$

For simplicity, the time weight W is omitted in this formula.

This simple model gives a strong indication of note boundaries at common intervals of music, which permits the identification of the current beat interval.

An illustration of the calculation of the beat probability vector can be seen in figure 3. It consists of the estimated audio feature (left), the estimation of probable beat and the updating of the running beat probability vector (right). The current beat interval is now found as the interval closest to the maximum in the beat probability vector. If no such interval exists, the maximum of the beat probability vector is used.

5 Evaluation

The beat estimation has been evaluated by comparing the beat per minute (BPM) output of the algorithm to a human estimate. The human estimate was found by tapping along while the musical piece was playing, and finding the mean time difference between taps.

To evaluate stability of the algorithm 10 pieces of popular and electronic music was randomly selected from a large music database. In all cases the algorithm gave a stable output throughout the piece, after a startup period of 1 to 60 seconds. The long startup period is due to the nature of the start of these pieces, i.e. non rhythmic music. In six of the cases the estimated BPM value matched the human estimate, while in the remaining four cases, the algorithm estimate was half that of the human estimate. The problem of not estimating the right multiple of BPM is reported elsewhere [3], however, it is worth noting that in the case of controlling the tempo of the music, it is of primary importance to have a stable output.

In addition, informal use of the system in real-time audio conducting [7], DJ beat matching and tempo control [6] has shown that the beat estimation is stable for a large variety of music styles.

6 Conclusions

This paper presents a complete system for the estimation of beat in music. The system consists of the calculation of an audio feature that has been selected from a large number of potential features. A number of error measures have been calculated, and the best feature has been found, together with the optimum threshold and block size, from the analysis of the error measures. The selected feature (high frequency content), is further enhanced in a beat probability vector. This vector, which keeps in memory the previous most likely intervals, renders an estimate of the current interval by the maximum of the beat interval probabilities.

The paper has presented several new features, a novel approach to the feature selection, and a versatile beat estimation that is both precise and immediate. It has been implemented in the DJ software Mixxx [14] and used in two well proven real-time music control systems: Conducting audio files [7] and DJ tempo control [6].

References

1. Goto, M., Muraoka, Y.: A real-time beat tracking system for audio signals. In: Proceedings of the International Computer Music Conference. (1995) 171–174
2. Goto, M., Muraoka, Y.: A real-time beat tracking for drumless audio signals: Chord change detection for musical decisions. Speech Communication **27** (1998) 311–335
3. Scheirer, E.D.: Tempo and beat analysis of acoustic musical signals. J. Acoust. Soc. Am. **103** (1998) 588–601
4. Dixon, S.: Automatic extraction of tempo and beat from expressive performances. Journal of New Music Research **30** (2001) 39–58
5. Desain, P.: A (de)composable theory of rhythm. Music Perception **9** (1992) 439–454
6. Andersen, T.H.: Mixxx: Towards novel DJ interfaces. Conference on New Interfaces for Musical Expression (NIME'03), Montreal (2003)
7. Murphy, D., Andersen, T.H., Jensen, K.: Conducting audio files via computer vision. In: Proceedings of the Gesture Workshop, Genova. (2003)
8. McAdams, S., Winsberg, S., Donnadieu, S., Soete, G.D., Krimphoff, J.: Perceptual scaling of synthesized musical timbres: Common dimensions, specificities, and latent subject classes. Psychological Research **58** (1995) 177–192
9. Jensen, K., Murphy, D.: Segmenting melodies into notes. In: Proceedings of the DSAGM, Copenhagen, Denmark. (2001)
10. Dixon, S., Goebl, W., Widmer, G.: Real time tracking and visualisation of musical expression. In: II International Conference on Music and Artificial Intelligence. Volume 12., Edinburgh, Scotland (2002) 58–68
11. Beauchamp, J.: Synthesis by spectral amplitude and "brightness" matching of analyzed musical instrument tones. Journal of the Acoustical Society of America **30** (1982)
12. Masri, P., Bateman, A.: Improved modelling of attack transient in music analysis-resynthesis. In: Proceedings of the International Computer Music Conference, Hong-Kong (1996) 100–104
13. Gordon, J.W.: The perceptual attack time of musical tones. J. Acoust. Soc. Am. **82** (1987)
14. Andersen, T.H., Andersen, K.H.: Mixxx. http://mixxx.sourceforge.net/ (2003)

Towards a General Architecture for Musical Archive Information Systems

Loretta Diana, Goffredo Haus, and Maurizio Longari

LIM-DICO University of Milan
via Comelico, 39 20135 Milano, Italy
{diana,longari}@dsi.unimi.it, haus@dico.unimi.it

Abstract. Musical Archive Information System (MAIS) is an integration of many research efforts developed at computer music laboratory of University of Milan. It is a system to organize, manage and utilize information of a heterogeneous set of music source material. Unstructured information is retrieved by non-traditional queries such as humming or playing of a melody (content queries). This work is a development of the previous information system developed in the context of project for Teatro Alla Scala. The improvements regard the use of XML format for the representation of Musical Work and the new user inter-faces for the navigation of musical source material.

1 Introduction

Musical Archive Information System (MAIS) is an integration of many research efforts developed at Laboratorio di Informatica Musicale (LIM). It is a system to organize, manage and utilize information of an heterogeneous set of music source material.

The main functionalities are database management and content data navigation. Database management allows the consultation of structured and unstructured multimedia information. Structured information is retrieved by traditional queries. Unstructured information is retrieved by non-traditional queries such as humming or playing of a melody (content queries). Content data navigation allows the non-sequential synchronized rendering of retrieved audio and score sources.

The framework of our information system is composed by three main modules: a multimedia database, an archive of musical files and a set of interfaces. The multimedia database has been implemented with object relational technology. Meta-data, links to archive material and content indexes are stored as MAIS database objects. Content indexes are musical features extracted from the source material in preprocessing phase. Furthermore, in this phase all the source material is analyzed for the automatic inclusion into the information system. The results are then included into a description XML file that works as an umbrella for the entire source files.

Other projects having similar data source like VARIATIONS project [6] at Indiana University and Online Music Recognition and Searching (OMRAS) project [17] addresses the need to represent symbolic information.

U.K. Wiil (Ed.): CMMR 2003, LNCS 2771, pp. 23–33, 2004.

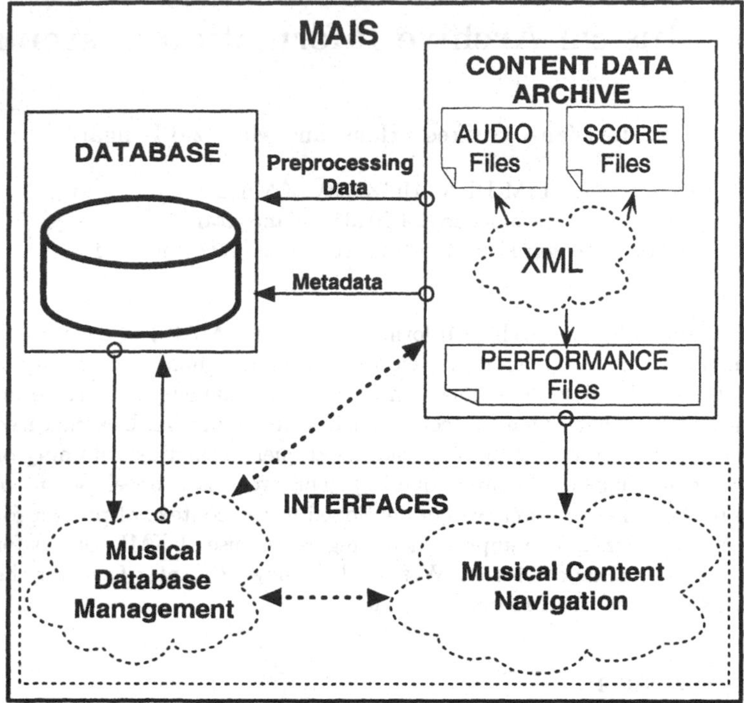

Fig. 1. MAIS overall architecture.

In the following sections we describe the framework of the system, the content data archive, the preprocessing and database architecture and the MAIS interfaces.

2 MAIS Framework

The framework of our information system is composed by three main modules as described in Figure 1. From the user point of view they are organized in levels of depth. At the lowest level there is the Content Data Archive that is the heterogeneous set of all source material. At the middle level is the database system with which the great amount of data actually present in the archive can be searched and retrieved. The front end of the system is a set of interfaces allowing the traditional and non-traditional queries and navigation of musical information.

2.1 Communication between Modules

The communication between modules is based on XML approach. We can distinguish among two types of XML messages: Commands and Data. Commands are represented with dotted arrow lines. While straight lines represent the passage of data between modules. We briefly describe the interaction of the modules.

When a new source material about a piece of music is introduced into the system it is firstly placed into the Content Data Archive (CDA). From the Musical Database Management (MDM) Module some meta-data is manually entered and some information is automatically extracted by means of a pre-processing phase. All of these data are stored into the database.

From the MDM module we can perform traditional queries about catalog information as well as content-based queries. Intelligent interfaces modalities allow whistling or humming a melody of the piece that we would retrieve.

When we have found the piece that we were searching for we can navigate into the source material by means of the Musical Content Navigation (MCM) Module.

3 MAIS: Content Data Archive

The Content Data Archive is composed of files containing several instances of musical information. They are grouped into Audio, Score, Performance and XML categories (Figure 1). These categories contain encoding of distinct aspects of musical information and constitute a big amount of data. To face the problem of information handling we encode descriptive and meta-data information in XML syntax.

The Score, Audio and Performance category contains the Source Material. The XML category contains information about a Music Work. A Music Work contains information about all its Source Material. The concept of Music Work is described in XML section below.

When new material is inserted into the archive it is analyzed and logged into the database and into the relative XML file (Figure 3a). Some of the formats of Source Material need a preprocessing phase for the extraction of musical features. Methods for elaboration are described in the relative sections.

A case study in which these concepts are fully used is the La Scala Project [10][13], developed at LIM [23]. In this project, the notion of Spine is the key element for the integration of scores (space dimension) with audio files (time dimension)[11].

In the following section, we briefly describe each category.

3.1 Score

In this category we group all the possible visual instances of a piece of music. We distinguish between two kinds of representation formats: notational and graphical.

Notational formats represent symbolic information (e.g. NIFF[16] and ENIGMA). This kind of information can be seen, from our point of view, as self-explaining one since it does not need elaboration to be interpreted but it needs only proprietary soft-ware translators.

Graphical formats are representation of scores images (for example TIFF, JPEG image or a PDF file). Before being included in the archive these sources must be pre-processed with Optical Music Recognition algorithm.

3.2 Audio

The Audio category contains musical audio information. Formats representing audio information can be subdivided in two categories: compressed or not compressed. Not compressed audio can be PCM/WAV, AIFF and Mu-Law. Compressed audio can be subdivided in lossy (e.g. MPEG and DOLBY) and lossless (ADPCM). Audio information is preprocessed at the moment of insertion into the archive. We perform a score-driven audio indexing [4], i.e. starting from symbolic information we search for event start times into audio file. It is possible to search for all musical events of an ensemble score or only for a single instrument part. These indexes are then stored in the XML file as meta-data information about the audio file.

3.3 Performance

Performance category embraces sub-symbolic formats such as MIDI files, Csound or SASL/SAOL (MPEG4). These formats can be seen as hybrid because they contain neither score nor pure audio information. They essentially encode information of what to play and how the sounds must be created.

3.4 XML

The main concepts of our structuring approach are layer-subdivision of musical information and space-time constructs (see [12]).

Each layer models a different degree of abstraction of music information. Essentially we distinguish among General, Structural, Music Logic, Graphic and Notational, Performance, and Audio layers (Figure 2a).

Music Logic is the core frame of our view of Symbolic Music Information. This information is composed by two fundamental concepts: the Spine and Logical Organized Symbols (Figure 2b).

Spine is a structure that relates time and spatial information. The measurement units are expressed in relative format. Thus, with such a structure, it is possible to move from some point in a Notational instance to the relative point in a Performance or Audio instance. Logical Organized Symbols are the common ground for the music content. Music symbols are represented in the XML syntax making music content explicit to applications and users.

The General layer is the place where meta-data information about the Music Work, Digital Rights and Copyright management is coded.

The Structural information layer is about the explicit description of music objects and their relationships from both the compositional and musicological point of view. The music information content within this layer does not contain explicit descriptions of time ordering and absolute time instances of music events. Causal relationships among chunks of music information and related transformations within the music score are described like they appear in the frame of compositional and analysis processes.

Notational layer groups the meta-data and relationships about Score Category of the archive. Information contained in this layer is tied to the spatial part of Spine structure.

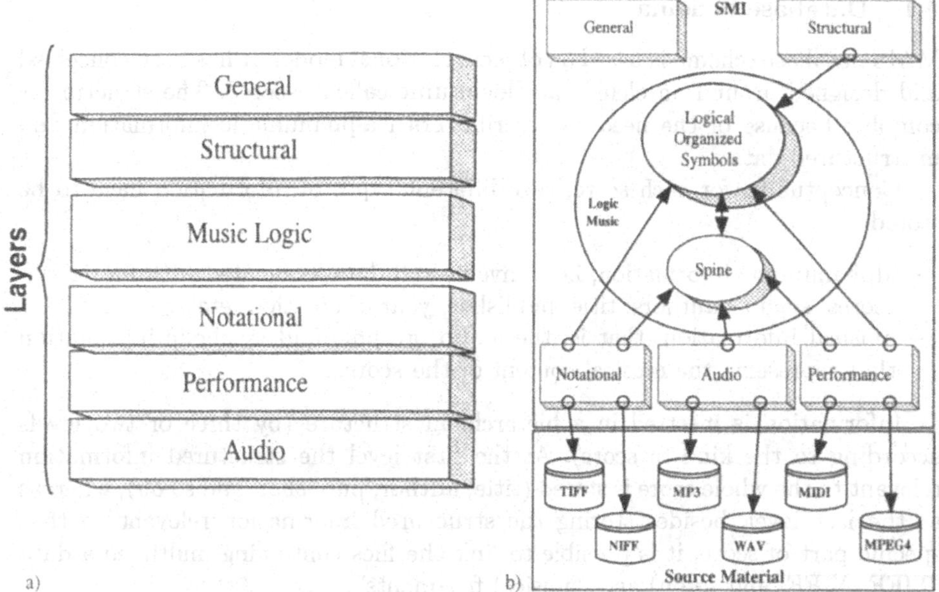

Fig. 2. Music work layers (a) and relations between them (b).

Audio layer groups the meta-data and relationships about Audio Category of the archive. This information relates to the time part of Spine structure. Note that this information is independent from the format in which the audio information is stored.

Performance layer groups the meta-data and relationships about Performance Category of the archive. These formats encode timing information in relative coordinates and thus they are naturally related to the Spine structure.

4 MAIS: Database Architecture

Multimedia data are inherently different from conventional ones [21], because information about their contents is not usually encoded with attributes as in traditional structured data -instead- multimedia data as audio items are typically unstructured. Specific methods to identify and represent content features and semantic structures of multimedia data are therefore needed [3]. Another distinguishing feature of multimedia data is their large storage requirements. One single image usually requires several Kbytes of storage, and a minute of audio can require several Mbytes. Relevant works in this area are the project developed by Ghias et al. at Cornell University [8], the one carried out at the University of Waikato in New Zealand [15].

In the following subsections we will discuss about database schema, musical information pre-processing and content-based queries.

4.1 Database Schema

MAIS database schema is based on object relational model. It has been conceived and designed around an elementary logic unit called "score". The structure is complex because of the need to describe both alphanumeric information and unstructured data.

Conceptually, for each score, two different types of information have to be stored:

- alphanumeric information, i.e. conventional data associated with music scores, such as authors, title, publisher, year of creation, etc...
- musical information, that is, the audio, graphic and symbolic information that represents the musical content of the score.

Information is inserted in a hierarchical structure (by three or two levels according to the kind of score). At the first level the structured information relevant to the whole score is stored (title, author, publisher, and so on), whereas at the last level, besides storing the structured information relevant to that specific part of score, it is possible to link the files containing multimedia data (TIFF, NIFF, and so on) and "musical fragments".

While encoding alphanumeric information is an easy task, storing musical information is a complex activity. Musical information of a score can be represented in several ways.

The scores are divided into their composing logic units, for example, the score of an opera is composed of acts and these further are divided in arias.

Furthermore, relationship is created between the score and the "musical fragments" extracted from the score itself that represent the fundamental themes, or source pat-terns (a sequence of limited number of notes), from which a great portion of the piece can be derived by means of musical function such as repetition, transposition, mirror inversion, retrogradation and other transformations and combinations. These notes are described by their pitch and duration.

These source patterns are needed for content-based query because the searching process is considerably more efficient on patterns than on a complete score. As well, maintaining pitch and duration for each note in a score for each score into the database is highly impractical due to the large space requirement. With this strategy we can record in the database the pitch and the duration only for the notes composing the source patterns [4]. Pitches and durations are extracted from the XML files stored into MAIS database during pre-processing phase.

4.2 Pre-processing Phase

When a score is acquired by MAIS, it passes through a pre-processing phase that consists of several steps to extract source pattern.

First, the XML file corresponding to the input score is generated. For the automatic acquisition of a music score and its conversion into XML files, Optical Music Recognition (OMR) is used (denoted as Digitization & Optical Music Recognition Module in Figure 3), which consists of three software modules. The first is the Digitizer Module, which transforms score sheets into TIFF files. The second is the OMR Pre-processor Module, to remove ambiguities from the TIFF files, slopes, bending, graphic noise are removed and other zones are processed to reduce or remove ambiguity, while problems due to the variable number of staves per page are solved. Thanks to this pre-processing phase, commercially available OMR products improve their efficiency by 30-50%. Finally, the OMR Recognizer Module extracts XML files from the enhanced TIFF files generated by the OMR Preprocessor Module.

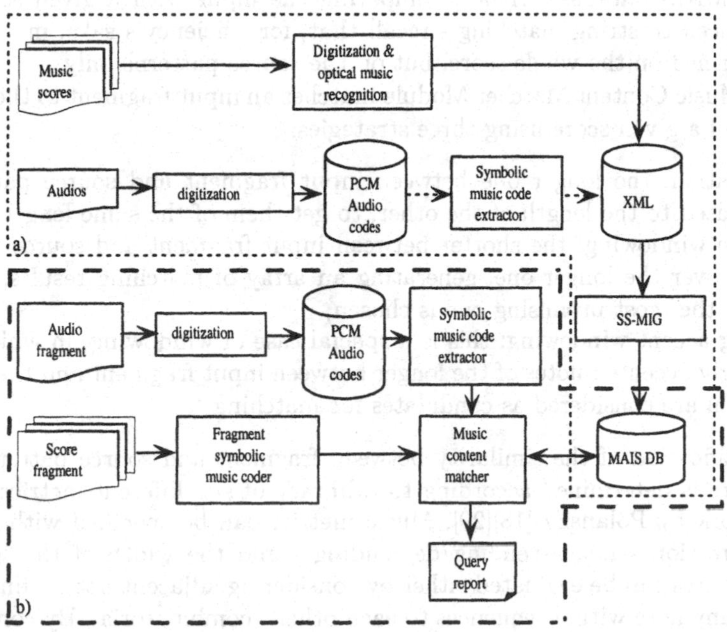

Fig. 3. Musical Data Management module.

The second step in this pre-processing phase is the extraction of source patterns. This activity is done by the Score Segmenter and Music Index Generator (SS-MIG) Module, shown in Figure 3a, software developed at LIM, capable of extracting source module patterns from XML files. Such patterns are extracted using a score segmentation algorithm [9].

In addition, it is possible to manually refine the output of the SS-MIG by adding, removing or modifying some of the patterns found by the SS-MIG.

4.3 Content Based Queries

The matching between the input and the scores stored into MAIS is performed in several steps, graphically illustrated in Figure 3b. The input can be either an audio file or a score fragment, played by the user on a keyboard or sung or whistled into a microphone connected to the computer [14]. From the audio files, note-like attributes are extracted by converting the input into a sequence of note-numbers, i.e., the concatenation of pitch and duration of each input note. Such step is performed by the Symbolic Music Code Extractor Module (Figure 3b). The conversion uses a Pitch-Tracking algorithm [19] based on conventional frequency-domain analysis. If the input is entered from a keyboard or it is a score fragment, conversion is not necessary and the sequence can be directly built. The Symbolic Music Code Extractor converts acoustic input into a score-like representation, using the same notation used to represent music scores into the MAIS database. Hence, comparing the input with a given score has been reduced to string matching - recall that, for efficiency's sake, matching is not performed on the whole score, but on the source patterns only.

The Music Content Matcher Module matches an input fragment to the source patterns of a given score using three strategies:

- truncation: the longer one between input fragment and source pattern is truncated to the length of the other, to get them of the same length;
- simple windowing: the shorter between input fragment and source pattern slides over the longer one, generating an array of matching results, among which the most promising one is chosen;
- strong accent windowing: this is a special case of windowing, in which only strongly accented notes of the longer between input fragment and the source pattern are considered as candidates for matching.

Quantification of the similarity between fragment and source patterns of a given score is determined according to a library of six different metrics, based on the work by Polansky [18][20]. Music metrics can be specified with respect to the direction - i.e., ascending/descending - and the width of the interval. These features can be evaluated either by considering adjacent notes - linear - or coupling any note within sequences to each other - combinatorial. By combining direction/width, linear/combinatorial, and ordering/non-ordering of sequences, eight metrics are obtained. In this library, combinatorial direction metrics are not considered, because they do not give any appreciable improvement to the search.

5 MAIS Interfaces

5.1 Musical Database Management (MDM) Module

One of modules provided by MAIS, that represents one of the most innovative features of the project, provides tools for efficient storage of musical data and for performing content-based queries on such data.

The overall architecture of the Musical Data Management module is illustrated in Figure 3. In the figure, dashed arrows denote features that are not yet implemented, but are currently under development. The module consists of two main environments: the Musical Storage Environment (Figure 3a) and the Musical Query Environment (Figure 3b). The musical storage environment has the purpose of representing musical information in the database, to make query by content efficient. The musical query environment provides methods to perform query by content on music scores, starting from a score or an audio fragment given as input - in the near future, the same functions will be provided for audio excerpts [7]. In the following subsections, both the above environments are discussed.

Storage Environment. The information is inserted through the forms that represent, also visually, the three hierarchical levels. At the first level the structured information relevant to the whole score is inserted (title, author, publisher, and so on), whereas at the last level, besides inserting the structured information relevant to that specific part of score, it is possible to link the files containing multimedia data (TIFF, MIDI, and so on) and to use the tools described before for their processing.

In addition, it is possible to manually refine the output of the SS-MIG by adding, removing or modifying some of the patterns found by the SS-MIG. To this end, the graphical environment is provided, which allows the user to insert patterns simply by clicking buttons on the keyboard. Once the patterns have been identified, either automatically or manually, information on pitch and duration of each pattern note is re-corded into the database.

Query Environment. The environment for querying music scores provides an integrated support for both standard and content-based queries. Content-based queries allow the retrieval of scores based on melodic similarity, a feature defined as melody retrieval. Melody retrieval allows a search for scores containing a given theme, or a sequence of notes similar to that theme.

Thanks to the musical query environment, a user can formulate both traditional queries such as: "Retrieve all the scores written by Mozart", or melodic queries such as: "Retrieve all the scores containing a particular sequence of notes (or a sequence of notes whose melodic content is similar to a given one)". Traditional and advanced selection criteria can be also combined. The user can submit queries on music scores by graphical interface.

Expert users of MAIS can define ad-hoc searching strategies for music retrieval (using the environment with Advanced matching mode), by combining metrics with different weights. Furthermore, the weight of pitch vs. duration in the matching process can be controlled. A set of operators can be used for retrieval to take into account logical closeness of related music fragments, such as when a fragment can be obtained from another with pitch transposition, retrogradation, mirror inversion, or a combination of the three. Non-experts are given a default combination of metrics (standard matching mode), so that musical contents are matched with a well-balanced combination. The returned music scores are then presented to the user, together with their similarity degree with the input. The user can then select a music score from the list, observe its graphical representation, while simultaneously playing the music.

5.2 Musical Content Navigation (MCN) Module

One of the key features of MAIS is the possibility of musical navigation by content. Score-driven audio features allow the synchronized audio-notation play-out of music. Musical Content Navigation (MCN) module takes an audio file (instance of a certain piece of music), a Notation file (for the moment we consider only MIDI files) and, using the audio features extracted in the insertion phase, renders a synchronized play-out of the music piece.

In Figure 1 we can see that MCN module is enabled by MDM module that carries out the search operations. Once the requested files are identified by MDM module, MCN module can access the Content Data Archive, retrieve the files and begin the play-out operation.

6 Conclusion

We have presented the overall architecture of MAIS. The system allows to organize, manage and utilize information of a heterogeneous set of music source material.

This work is a development of the previous information system developed in the context of project for Teatro Alla Scala[5]. The improvements regard the use of XML format for the representation of Musical Work[12] and the new user interfaces for the navigation of Source Material[4].

Further work must be done on the preprocessing phase such as symbolic feature extraction from complex audio. XML music format representation is still under development in the context of standardization process of MAXWG[22].

Acknowledgements

The authors wish to acknowledge the partial support of this project by the Italian National Research Council, in the frame of the research program "Methodologies, te-chniques, and computer tools for the preservation, the structural organization, and the intelligent query of musical audio archives stored on heterogeneous magnetic media", Finalized Project "Cultural Heritage", Subproject 3, Topic 3.2, Subtopic 3.2.2, Target 3.2.1.

This work has been made possible by the effort of researchers and graduate stu-dents of LIM.

References

1. AA.VV.: Standard MIDI File 1.0. The International MIDI Association, (1988).
2. AA.VV.: MIDI 1.0 Detailed Specification. The International MIDI Association, (1990).
3. Bertino, E., Catania B., Ferrari, E.: Multimedia IR: Models and Languages. Modern Infor-mation Retrieval, Addison-Wesley (1999)
4. D'Onofrio, A.: Methods for Integrated Timed Audio and Textual Digital Music Processing. Master Thesis, Computer Science Department, University of Milan (1999)

5. Diana, L., Ferrari, E., Haus, G.: Saving the Multimedia Musical Heritage of Teatro alla Scala for Querying in a Web-Oriented Environment. Proc. of the IEEE 2001 Wedelmusic Conference, Florence (2001)
6. Dunn, J.W., Mayer, C.A. "VARIATIONS: A Digital Music Library System at Indiana Uni-versity" Proceeedings of the Fourth ACM Conference on Digital Libraries, Berkeley, CA, Aug 1999, pp. 12-19
7. Frazzini, G., Haus, G., Pollastri, E.: Cross Automatic Indexing of Score and Audio Sources: Approaches for Music Archive Applications. Proc. of the ACM SIGIR '99 Music Informa-tion Retrieval Workshop, Berkeley, CA, USA (1999)
8. Ghias, A., Logan, J., Chamberlin, D., Smith, B.C.: Query by Humming: Musical Information Retrieval in an Audio Database. In Proc. of the ACM Multimedia Conference, San Fran-cisco, CA, November (1995) 5-9
9. Guagnini, S.: Strategies and Tools for the Automatic Segmentation of Music. Master Thesis, Computer Science Department, University of Milano (1998)
10. Haus, G.: Rescuing La Scala's Audio Archives. IEEE Computer, Vol. 31(3), IEEE CS Press, Washington (1998) 88-89
11. Haus, G., Longari, M.: Coding Music Information within a Multimedia Database by an Inte-grated Description Environment. Atti del XII Colloquio di Informatica Musicale, AIMI/Università di Udine, Gorizia (1998)
12. Haus, G., Longari, M: Towards a Symbolic/Time-Based Music language based on XML. IEEE Proc. of MAX2002, Milan (2002)
13. Haus, G., Pelegrin, M. L.: Music Processing Technologies for rescuing Music Archives at Teatro alla Scala and Bolshoi Theatre. Journal of New Music Research, Swets & Zeilinger, Amsterdam (2001)
14. Haus, G., Pollastri, E.: A Multimodal Framework for Music Inputs. In Proc. of the ACM Multimedia, ACM Press, Los Angeles (2000)
15. McNab, R., Smith, L.A., Witten, I.A.: Towards the Digital Music Library: Tune Retrieval from Acoustic Input. In Proc. of the ACM Digital Library Conference (1996) 11-18
16. Mounce, S.: NIFF 6a. Available at: http://www.student.brad.ac.uksrmounce/niff.html
17. On Line Music Recognition And Searching (OMRAS) http://www.omras.org
18. Polansky, L.: Morphological Metrics. Journal of New Music Research, vol. 25, pages 289-368, Swets & Zeilinger Publ., Amsterdam, 1996
19. Pollastri, E.: Memory-Retrieval based on Pitch-Tracking and String-Matching Methods. In Proc. of the 12th Colloquium on Musical Informatics, Gorizia, Italy (1998) 151-154
20. Salterio, R.: Methods for Music Contents Searching within Symbolic Score Databases. Mas-ter Thesis, Computer Science Department, University of Milan (1998)
21. Subrahmanian, V.S.: Principles of Multimedia Database Systems. Morgan-Kaufann (1997)
22. http://www.lim.dico.unimi.it/IEEE/XML.html
23. http://www.lim.dico.unimi.it

The Study of Musical Scales in Central Africa: The Use of Interactive Experimental Methods

Fabrice Marandola

LMS (UMR 8099 CNRS-Paris V, France)
Bât.D, 7 rue Guy Môquet
94801 Villejuif, France
marandol@vjf.cnrs.fr

Abstract. In the oral traditional cultures of Central Africa, the rules which underline the musical system are rarely verbalised: abstract concepts like "scale", "degree" or "interval", are not just non-verbalised, they are practically unverbalisable. Thus the study of musical scales requires the use of interactive experimental methods. For two years, a research-team leaded by Simha Arom (in collaboration with acousticians of Ircam-Paris) studies the untempered scales used by the Bedzan Pygmies in their contrapuntal songs and the Ouldeme in their hocket instrumental polyphony. The principle of the methods that we have developed is to make the musicians actors in the experiment, able to react immediately to the proposals of the investigators and to provide modifications of them, directly or indirectly.

1 Introduction

In the oral traditional cultures of Central Africa, the rules which underline the musical system are rarely verbalised: abstract concepts like "scale", "degree" or "interval", are not just non-verbalised, they are practically *unverbalisable*; there is indeed conception, but not conceptualisation. Thus the study of musical scales requires the use of *interactive experimental methods*. This seems the only way to catch the principles on which they are based on.

Pioneering work in the field of instrumental music — the tuning of xylophones — has been conducted by Simha Arom since 1989, in Central Africa and in Indonesia [[1,2]]. The recent development of most sophisticated acoustical computerised equipment has enabled us to enlarge the framework of interactive experimentation by adapting it also to vocal music, be it monodic or polyphonic. For two years (2000-2002), a research-team leaded by Simha Arom studied the untempered scales used by the Bedzan Pygmies in their contrapuntal songs and by the Ouldeme in their hocket instrumental polyphony[1].

The scales of these two ethnic groups are not only non-tempered, but also present some remarkable properties. In Bedzan Pygmies vocal polyphony for example, the same

[1] This research was supported by French Ministry for Research, Action Concertée Incitative *Cognitique*, project A 108 "Conception and perception of musical scales in traditional oral cultures: the case of Ouldeme and Bedzan Pygmies of Cameroon", and has been conducted in a collaboration between LMS (UMR 8099 CNRS/Paris V) and Ircam.

U.K. Wiil (Ed.): CMMR 2003, LNCS 2771, pp. 34–41, 2004.

piece presents a wide mobility of the tuning of the scale degrees from one version to another, or it can be sung either with a *tetratonic* or a *pentatonic* scale. In Ouldeme instrumental polyphony, the pentatonic scale which is observed in the low register is not reproduced identically in the higher register, and the octave seems not to be the frame which structures the entire scale [5].

The principle of the methods that we have developed is to make the musicians actors in the experiment, able to react immediately to the proposals of the investigators and to provide modifications of them, directly or indirectly. The goal is to progressively arrive at a model of the scale system used by the community, in other words, to reveal *the collective mental representation that the holders of a tradition have of their musical scale.*

This paper proposes to describe two methods which have been applied with Bedzan and Ouldeme musicians on their own traditional repertoires.

2 Hocket Instrumental Polyphonies of Ouldeme

Ouldeme music is composed of several flute orchestras. Only one of them is reserved for women who play the *azèlèn* flutes. These flutes are a set of ten reed tubes with no finger holes. The flutes are grouped in pairs (although the two instruments which make up the pair are not tied together) and divided among five instrumentalists. When playing, each musician inserts vocal sounds, which generally reproduce the instrumental pitches played by the contiguous flutes, between two instrumental sounds. These vocal sounds thus add to the complexity of the hocket polyphony already produced by the ensemble of ten instruments. The range of the ensemble covers almost two octaves. However, the pentatonic scale observed in the lowest register is not reproduced identically in the higher register.

Unlike the study about xylophones scales - in which researchers were working with one or almost two instrumentalists - here we had to work with five instrumentalists playing each with two flutes. In collaboration with acousticians of Ircam (Paris)[2], we built a set of electronic flutes managed by a MIDI-system. Each flute was equipped with a breath-controller - driving a physical-model synthesis of sound[3] - and two buttons - pressed by the flutists themselves to change the pitch of the sound. These two buttons simulated the addition of water or its pouring off, which is the traditional way to tune the flutes.

The whole process was controlled in real-time with a specific application developed by Manuel Poletti with Max/MSP. There were three interfaces to check Midi I/O, to control the physical-model synthesis of sound and to drive the experiments. This last interface (see below) was divided in three sections: 1/ "Mixing", to mix the sounds of the ten flutes, 2/ "Tuning", to modify the step of pitch modification for each pressure on a button of the flutes, and to check in real-time the changes made by all the instrumentalists to tune their flutes, and 3/ "Midi-sequencer", to record all the values and to play back the sequences of playing.

[2] René Caussé, Claire Ségouffin, Christophe Vergez (Ircam, Paris), Patricio de la Cuadra (Stanford, USA).

[3] For more informations about the physical-model synthesis of sound, see [3, 4, 6].

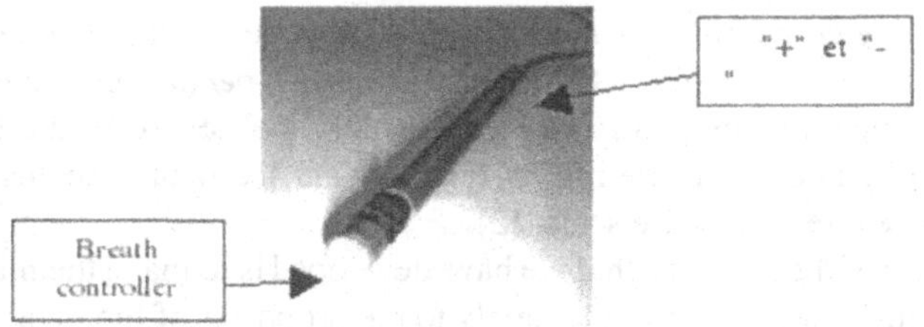

Fig. 1. Bakelite flute with breath-controller and buttons for the tuning of the pitch (photography: René Caussé, Ircam)

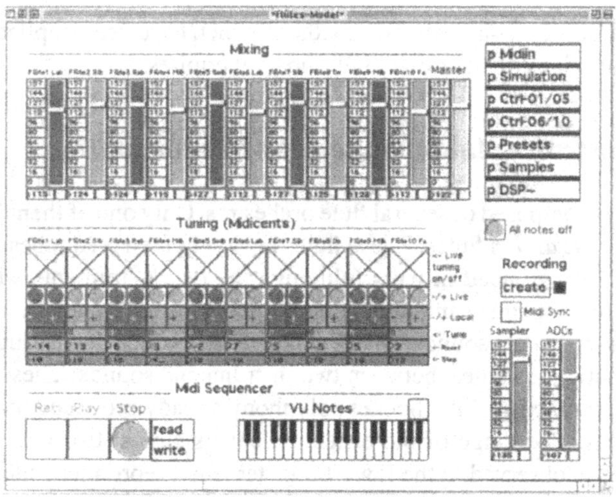

Fig. 2. Interface for driving the experiments

During four days, the Ouldeme women took part to the experiments. Very quickly, they felt comfortable with these strange flutes in a such manner that, the third day, they were discussing like "Push twice" or "Add one more" (i.e. "push twice on this button" or "push again on the upper button"). It means that they were able to determine, in an abstract manner and with a step of 10 *cents* for each pressure, the distance between the pitch like it was tuned from what it should be for her. Without this experimental method, we would never have such precise and convincing results.

2.1 Vocal Polyphonies of Bedzan Pygmies

Bedzan music is mostly composed of four-part polyphonic songs, accompanied by one or several drums and wicker rattles, and sometimes hand clapping. The performance of a song lets all members of an encampment take part; the parts of the song correspond to four complementary registers: *nkw? b?nkn*, the "part of adult (men)", *nkw? by*, the "part of (adult) women", *nkw? bembeban*, the "part of young men" and *nkw? bw?s*, the "part of the little ones (children)". The songs are started by a soloist, male or female, followed

Fig. 3. Example of Bedzan' song (extract = one cycle)

by the four constituent voices of the chorus entering by superposition or juxtaposition (all voices together or soloist/chorus alternation), the latter technique apparently linked to more recent songs. In most cases, only the solo part which begins the song has words. The rest of the song is composed of vowels which allow the singers to freely perform different variations. The Bedzan music is cyclical, based on the repetition of a musical figure which is understood by all and which can be varied at the singers' will as the song goes on, without loosing its identity. The participants thus have a degree of flexibility which is further extended by the possibility of switching parts at certain precise moments in the cycle.

Our research method — used *in situ*, in Cameroon — requires software for sound signal analysis-synthesis, for formalisation and analysis of the musical language, and equipment for multi-track digital audio recording.

The first phase involves recording simultaneously — but also, thanks to a multi-track system, separately — all of the parts of a polyphonic song: each singer wears a head-worn microphone that records only his/her voice, allowing complete freedom of movement [4]. Each part, digitised in real time during the recording, can then be analysed separately.

The second step is the spectral analysis (with *AudioSculpt* software) of each voice part in order to measure the fundamental frequencies of the various notes. The dispersion of the degrees is then evaluated based on these fundamental frequencies with a specific computer application (*Scala* [5]) developed in the *OpenMusic* software: various curves and graphs allow to visualise the behaviour of each voice in its temporal

[4] In order to facilitate the analysis work done with the computers, the singers are asked to sing simple versions. This initially eliminates the problem of accumulation of melodic variations: the only variations that interest us at this stage are the interval variations between the same degrees of a scale.

[5] To be published in Forum Ircam 10-03 (CD-Rom) [7].

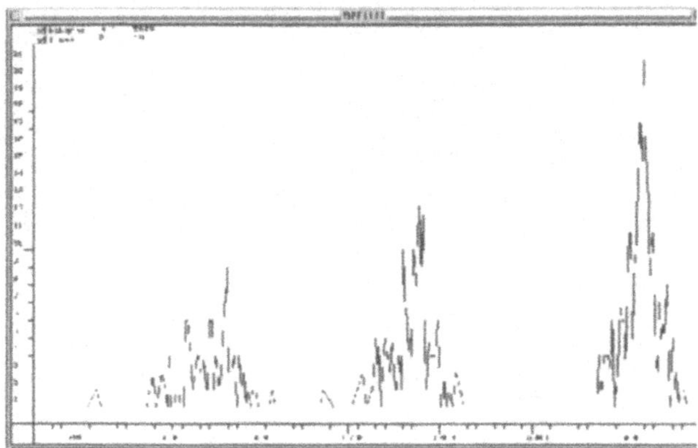

Fig. 4. Visualisation of the pitch dispersion for each degree of one vocal part

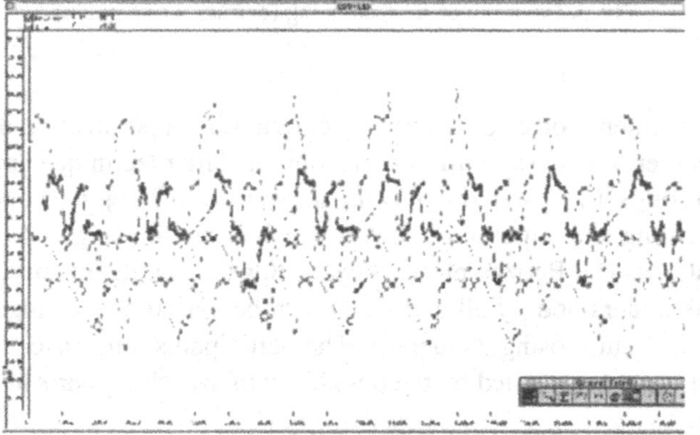

Fig. 5. Visualisation of the pitch dispersion for each degree of one vocal part

continuity, and to evaluate the field of dispersion of the degrees; these schemata can be displayed in both monodic and polyphonic form. The measured observation of the pitch fluctuations allow for the formulation of *hypotheses* concerning the size of the intervals, their distribution and the determination of *classes of intervals* in the musical scale, in the form of models applicable to the entire polyphony.

All of the parts are then modified – as a function of these various hypotheses – to reconstitute the polyphony, in accordance with the size of the intervals which have just been modelled, without modifying the timbre of the singers' voices nor the metric and rhythmic structure of the polyphony.

The next step involves submitting these various hypotheses for listening by the musicians, who perfectly recognise the piece and also the voices of the singers because their timbre remains unchanged; in this way, they are led to focus on the pitch parameter. They accept or reject each of the models one by one, thereby guiding the investigator in the development of new proposals. When one of the hypotheses is refused, the computer system allows for segmenting of the melodic contour in order to precisely locate the error and to more precisely check with the musicians regarding the reason for their refusal. The multi-track recording system offers the possibility of listening to the parts either in isolation or in polyphony; according to the musicians' suggestions, corrections can thus be made partially to one of the parts of the polyphony or may be applied to all of the parts present. In the same way, it is also possible to record separately the parts of one or several musicians while allowing all of them to simultaneously hear (with headphones) all of the complementary "computer's" parts, in order to verify that singers can *really* use the scale proposed or to correct the parts that have inadequate scale proposals.

The goal is to progressively arrive at a model of the scale system used by the community: the observation of the stability of the intonation behaviour of the singers allow us, by delimiting the field of dispersion of the degrees, to get as close as possible to the implicit cultural model which underlies their vocal practice.

3 Results

We applied these two methods in Paris for Ouldeme flutes, and in the forest of Cameroon for the Bedzan Pygmies.

3.1 Bedzan Music

The conception of the scales is based on *structuring intervals of less than an octave*. These are trichords (interval that itself contains two linked intervals) or *tetrachords* (interval containing three linked intervals). There can be relatively wide margins of realisation for each of these framework-intervals, as for the intervals that make them up. These margins are however limited by a triple constraint related to 1) the upper and lower limits accepted for the trichords and tetrachords (respectively 500/600 cents and 600/750 cents), 2) the upper and lower limits of the octave 1100/1250 cents), formed by the juxtaposition of the framework-intervals and 3) by the minimum and maximum values that the intervals that make up these framework-intervals can take.

These fundamental rules are supplemented by the possibility, for certain pieces, of a specific type of organisation of the framework-intervals with respect to the *central core* formed by the degrees sung by the various parts of the polyphony.

The Bedzan's conceptual model is dynamic. Based on a limited number of rules, it applies a set of *reciprocal constraints*, the equilibrium of which must be respected in order to ensure the continuity of the system. It is probably the very simplicity of the system that makes it simultaneously so free and so complex, ensuring easy transmission from generation to generation.

3.2 Ouldeme Flutes

The scale system of the Ouldémé flute ensemble has some points in common with that of the Bedzan. It is governed by interval relationships subject to *reciprocal constraints*, the trichords and tetrachords playing a leading role in the conception of the system.

The scale revealed does have many specific features however in that:

- there is *no complementarity* between two consecutive framework-intervals ending at the octave;
- *the octave is therefore not the reference interval* within which the scale is organised. On the contrary, there is a translation of a smaller structure organised according to a single tetrachord, this configuration being translated from one pair of flutes to another.
- nor is there a *central core.*

The scale system is based on a single type of interval for which the realisation margins are extremely wide (ranging from about 120 to 320 *cents*). This interval is however subject to a double constraint involving the limits of the trichord and the tetrachord. It is thus a system ruled by a single type of interval size — to determine the space between two linked degrees —, and also by the combinatory of *three types of intervals* (simple, trichord, tetrachord intervals) which this single interval contributes to construct — to define the global structure.

It should be noted that while each of them can have quite different realisations, these three intervals are clearly in contrast within the sound *continuum*, because they have no common "boundary": there can be no ambiguity, neither between the simple interval and the trichord, because their respective upper and lower limits are 50 *cents* apart, nor between the trichord and the tetrachord, separated by 30 *cents*.

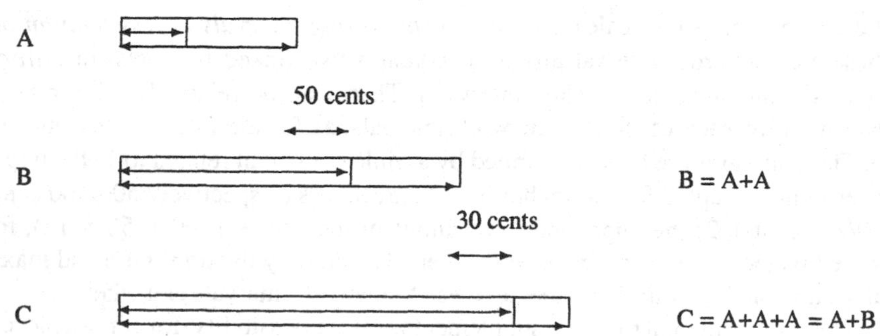

Fig. 6. Stucturing intervals of the Ouldeme scale

4 Conclusion

Only such methods allow us to understand how the musicians *conceive* of their musical system, despite the fact that the foundations of the system are rather abstract and the result of a long process of music learning, through impregnation and imitation, starting from early childhood. By overcoming the absence of verbalisation, they place researchers and musicians on an equal footing around the musical material. They give to the investigator the opportunity to receive unequivocal answers to his questions, in such a manner that he may have gradually an in-depth understanding of how the natives conceive of their musical system.

References

1. Arom, S. & Voisin, F.: Theory and Technology in African Music. *In: Stone, R. (ed.): The Garland Encyclopaedia of World Music, Vol. 1 (Africa). Garland, New York* (1998) 254-270
2. Arom, S., Voisin, F. & Léothaud, G.: Experimental Ethnomusicology: An Interactive Approach to the Study of Musical Scales. *In: Deliège, I., Sloboda, J. (eds.): Perception and Cognition of Music. Hove (U.K.), Erlbaum, Taylor & Francis* (1997) 3-30
3. Caussé R., de la Cuadra, P., Fabre, B., Ségoufin, C., Vergez, C.: Developing experimental techniques and physical modeling for ethnomusicology project on Ouldémé flutes. *In: Proceedings of the Forum Acusticum (European and Japanese Symposium on Acoustics). Seville, CD-Rom MUS-06-014* (2002)
4. Cuadra, P. de la, Vergez, C., Caussé, R.: Use of physical-model synthesis for developing experimental techniques in ethnomusicology-The case of the Ouldémé flute. *In: Proceedings of the International Computer Music Conference. Göteborg, Suède* (2002) 53-56
5. Fernando, N.: A propos du statut de l'octave dans un système pentatonique du Nord Cameroun. *In: MusicæScientiæ, Discussion Forum n°1. ESCOM, Liège* (2000) 83-89
6. Lamoine, J.: Etude des propriétés d'un jet turbulent soumis à un champ acoustique transversal, Stage de Maîtrise de Mécanique.*Université Paris XI, Orsay* (2001)
7. Marandola, F.: Scala. *In: Forum Ircam 10-03. CD-Rom, Ircam, Paris*(2003).

Evolving Automatically High-Level Music Descriptors from Acoustic Signals

François Pachet and Aymeric Zils

Sony CSL Paris
6 rue Amyot, 75005 Paris, France
{pachet,zils}@csl.sony.fr

Abstract. High-Level music descriptors are key ingredients for music information retrieval systems. Although there is a long tradition in extracting information from acoustic signals, the field of music information extraction is largely heuristic in nature. We present here a heuristic-based generic approach for extracting automatically high-level music descriptors from acoustic signals. This approach is based on Genetic Programming, that is used to build extraction functions as compositions of basic mathematical and signal processing operators. The search is guided by specialized heuristics that embody knowledge about the signal processing functions built by the system. Signal processing patterns are used in order to control the general function extraction methods. Rewriting rules are introduced to simplify overly complex expressions. In addition, a caching system further reduces the computing cost of each cycle. In this paper, we describe the overall system and compare its results against traditional approaches in musical feature extraction à la Mpeg7.

1 Introduction and Motivations

The exploding field of Music Information Retrieval has recently created extra pressure to the community of audio signal processing, for extracting automatically high level music descriptors. Indeed, current systems propose users with millions of music titles (e.g. the peer-to-peer systems such as Kazaa) and query functions limited usually to string matching on title names. The natural extension of these systems is content-based access, i.e. the possibility to access music titles based on their actual content, rather than on file names. Existing systems today are mostly based on editorial information (e.g. Kazaa), or metadata which is entered manually, either by pools of experts (e.g. All Music Guide) or in a collaborative manner (e.g. the MoodLogic). Because these methods are costly and do not allow scale up, the issue of extracting automatically high-level features from the acoustic signals is key to the success of online music access systems.

Extracting automatically content from music titles is a long story. Many attempts have been made to identify dimensions of music that are perceptually relevant and can be extracted automatically. One of the most known is tempo or beat. Beat is a very important dimension of music that makes sense to any listener. Scheirer introduced a beat tracking system that successfully computes the beat of music signals with good accuracy ([1]).

U.K. Wiil (Ed.): CMMR 2003, LNCS 2771, pp. 42–53, 2004.

There are, however, many other dimensions of music that are perceptually relevant, and that could be extracted from the signal. For instance, the presence of voice in a music title, i.e. the distinction between instrumentals and songs is an important characteristic of a title. Another example is the perceived intensity. It makes sense to extract the subjective impression of energy that music titles convey, independently of the RMS volume level: with the same volume, a Hard-rock music title conveys more energy than, says, an acoustic guitar ballad with a soft voice.There are many such dimensions of music that are within reach of signal processing: differentiate between "live" and studio recording, recognize typical musical genres such as military music, infer the danceability of a song, etc... Yet these information are difficult to extract automatically, because music signals are usually highly complex, polyphonic in nature, and incorporate characteristics that are still poorly understood and modeled, such as transients, inharmonicity, percussive sounds, or effects such as reverberation.

1.1 Combining Low-Level Descriptors (LLD)

Feature extraction consists in finding characteristics of acoustic signals that map correctly with values obtained from perceptive tests. In this context, the traditional approach in designing an extractor for a given descriptor is the following (see, e.g. [1–3]):

Firstly, perceptive values are associated to a set signal of from a reference database. These values can be obvious (Presence of singing voice), or can require to conduct perceptive tests (Evaluation of the global energy of music titles): humans are asked to enter a value for a given descriptor, and then statistical analysis is applied, to find the average values, considered thereafter as a grounded truth.

Secondly, several characteristics of the associated audio signals are computed. A typical reference for audio characteristics is the Mpeg7 standardization process ([5]), that proposes a battery of LLD for describing basic characteristics of audio signals. The purpose of Mpeg7 is not to solve the problem of extracting high level descriptors, but rather to propose a basis and a format to design such descriptors.

Eventually, the most relevant LLDs are combined in order to provide an optimal extractor for the descriptor.

1.2 Two Illustrative Examples

We illustrate here descriptor extraction, using the standard approach on two music description problems, that are relevant for music information retrieval, objective, and difficult to extract automatically.

The first problem consists in assessing the perception of energy in music titles. This descriptor yields from the intuitive need for differentiating between energetic music, for instance Hard Rock music with screaming voices and saturated guitar, from quiet music, such as Folk ballads with acoustic guitar, independently of the actual volume of the music. We have conducted a series of perceptive tests on two databases of 200 titles each. For each title, we asked the listeners to rate the "energy conveyed" from "Very Low" to "Very High". We got 4500 results, corresponding to 10 - 12 answers for each title. The analysis showed that for 98% of the titles, the standard deviation of the answers was less than the distance between 2 energy categories, so the perception of subjective energy is relatively consensual, and it makes sense to extract this information from the signal. The energy is a float number normalized between 0 (no energy) and 1(maximum energy).

The second problem consists in discriminating between instrumental music and songs, i.e. to detect singing voice. The technical problem of discriminating singing voice with speech in from a complex signal is known to be difficult ([6,7]) and remains largely open. No experiment were performed for this descriptor as the values are obvious to assess. The presence of voice is a Boolean value 0 (instrumental) or 1(song).

1.3 Results Using Basic LLD

We present here the results obtained on our two problems, using the standard approach sketched above. More precisely, the palette of LLD used for our experiments consisted of 30 LLD, obtained as Mean and Variance of:

Amplitude Signal, Amplitude Fft, High freq content, Max spectral freq, Ratio high freq, RMS, Spectral Centroid, Spectral Decrease, Spectral Flatness, Spectral Kurtosis, Spectral Roll Off, Spectral Skewness, Spectral Spread, Total Energy, Zero Crossing Rate. The method consisted in finding the linear combination of LLD that best matches the perceptive results.

First, the optimal combination is computed on each learning database: for the Global Energy problem, the regression consists in minimizing the average model error compared to the perceptive results; for the Instrumental/Song problem, the classification consists in maximizing the discrimination and finding a threshold to separate the 2 classes.

Then the combination is tested on each test database: for the Global Energy problem, the evaluation consists in computing the average model error compared to the perceptive results; for the Instrumental/Song problem, the evaluation consists in computing the recognition rate.

A cross-validation ensures the consistency of the method. The final results presented in Table 1 are the mean results of the cross-validations with their uncertainty:

Table 1. Results obtained using basic Mpeg7 LLD combination on two high-level description problems

	Subjective Energy (Model Error)	*Presence of Voice* (Recognition Rate)
Best Feature	16.87% +- 1.48%	61.0% +-10.21%
Best Combination	12.13% +-1.97%	63.0% +-11.35%

The success of the standard approach is dependent on the nature and quality of the basic signal extractors in the original palette. Mpeg7 provides some interesting descriptors, in particular in the field of spectral audio, but to extract complex, high-level musical descriptors, the features have to be much more specific. The next section gives a typical example of the failure of such an approach on a simple problem.

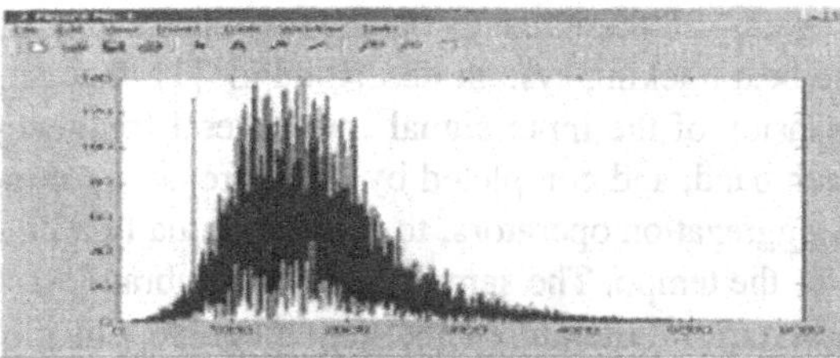

Fig. 1. Spectrum of a 650Hz sinus mixed with 1000-2000Hz colored noise

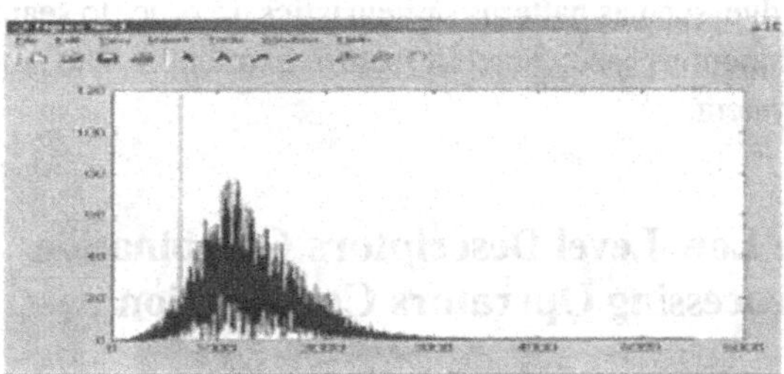

Fig. 2. Spectrum of a 650Hz sinus mixed with 1000-2000Hz colored noise, pre- filtered by a 1000Hz Low-Pass Filter

1.4 Limitations of the Traditional Method

Example: Sinus + Colored Noise. Let us consider the simple problem of detecting a sinus wave in a given frequency range (say 0-1000Hz) mixed with a powerful colored noise in another frequency range (1000-2000Hz). As the colored noise is the most predominant characteristic of the signal, traditional features will focus on it and are therefore unable to detect our target sinusoid. For instance, when we look at the spectrum of a 650Hz sinus mixed with a 1000-2000Hz colored noise (Fig.1), the peak of the sinus is visible but not predominant, and is thus very hard to extract automatically.

Of course, this problem is easy to solve by hand, by applying a pre-filtering that cuts off the frequencies of the colored noise, so that the sinus emerges from the spectrum. As seen on Fig.2, the sinus peak emerges when the signal is low-pass filtered, and is thus very easy to extract automatically.

Motivations for EDS. This basic example shows that the combination of basic LLD does not cover a function space wide enough to find specialized extractors. It is not the case that any high-level descriptor can be obtained by some linear combination of basic LLD. So an automatic system that produces extractors should be able to search in a larger and more complex function space, as signal processing experts normally do. The variations concern not only the actual operators used in the whole process, but also their parameters, and the possible "in-between" process such as filters, peak extractors, etc... that can be inserted to improve the accuracy of the extractor.

Although there is no known general paradigm for extracting relevant descriptors, the design of high-level extractors usually follow regular patterns. One of them consists in filtering the signal, splitting it into frames, applying specific treatments to each segments, then aggregating all these results back to produce a single value. This is typically the case of the beat tracking system described in [1], that can schematically be described as an expansion of the input signal into several frequency bands, followed by a treatment of each band, and completed by an aggregation of the resulting coefficients using various aggregation operators, to yield eventually a float representing (or strongly correlated to) the tempo. The same applies to timbral descriptors proposed in the music information retrieval literature ([8,9]). Of course, this global scheme of expansion/reduction is under specified, and a virtually infinite number of such schemes can be searched. Our motivation is to design a system that is able to use a given signal-processing knowledge, such as patterns or heuristics, in order to searches automatically signal processing functions specialized in feature extraction. The next Section presents the design of the system.

2 EDS: From Low-Level Descriptors Combination to Signal Processing Operators Composition

The key idea of our approach is to substitute the combination of basic LLD by the composition of operators. Our Extraction Discovery System (called EDS) aims at composing automatically operators to discover signal processing functions that are optimal for a given descriptor extraction task.

The core search engine of EDS is based on genetic programming, a well-known technique for exploring functions spaces [10]. The genetic programming engine automatically composes operators to build functions. Each function is given a fitness value which represents how well the function performs to extract a given descriptor; this is typically the correlation between the function values and the perceptive values. The evaluation of a function is therefore very costly, as it involves complex signal processing on whole audio databases. To guide the search, a set of *heuristics* are introduced, to control the creation of functions, as well as *rewriting rules* that simplify functions before their evaluation. This section presents EDS design principles.

2.1 Representation of Basic Signal Processing Operators

Each operator is defined by its name, its output type, and an executable program, which evaluate the function once it is instantiated. In EDS, these programs are written and compiled in Matlab. EDS functions include constants, mathematical operators such as mean or variance, signal processing operators, temporal such as correlation, or spectral such as FFT or filters. To account for the specificity of audio extraction, we introduced operators to implement the global extraction schemes. For instance, the *Split* operator splits a signal into frames, an operation that is routinely performed when a given treatment has to be made on successive portions of the signal.

Functions are built by composing these operators, each function containing at least one argument labeled *InSignal*, which is instantiated with a real audio signal before evaluation. Fig.3 shows an example of tree representation for a function that is a composition of basic operators (FFT, Derivation, Correlation, Max):

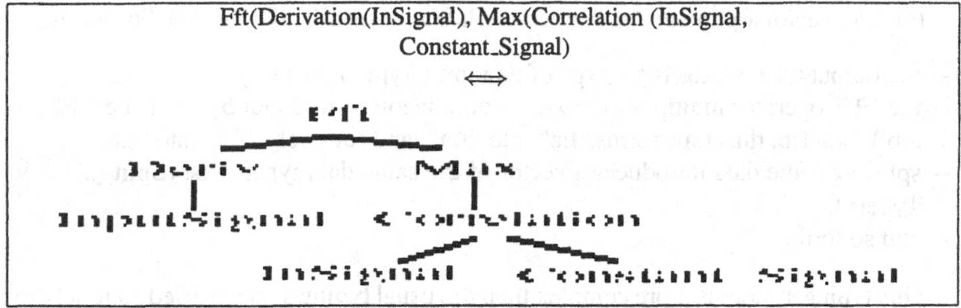

Fig. 3. Tree-like representation of a signal processing function

2.2 Data and Operators Types

The need for typing is well-known in Genetic Programming, to ensure that the functions generated are at least syntactically correct. Different type systems have been proposed for GP, such as strong typing ([11]), that mainly differentiate between the "programming" types of the inputs and outputs of functions.

In our context, the difference between programming such as types floats, vectors, or matrix, is superficial. For example, the operator "Abs" (absolute value) can be applied on a float, a vector, etc. This homogenous view of values types yields simplicity in the programming code, that we need to retain. However, we need to distinguish functions in terms of the "physical dimensions" they manipulate. Indeed, audio signals and spectrum can be seen both as vectors of floats from the usual typing perspective, but they are different in their physical dimensions: a signal is a time to amplitude function, while a spectrum associates frequency to amplitude. Our typing system, based on the following constructs, has to represent this difference and more generally physical dimensions, to ensure that our resulting functions make sense in terms of signal theory.

Atomic Types, Functions, Vectors. Types can be either atomic dimensions, which are of 3 sorts: time, notated "t"", frequency, notated "f", and amplitudes notated "a". These 3 types are the basis to build more complex types: functions and vectors.

Functions are objects that map one dimension to another. Their type is represented using the ":" notation, which differentiates between the x and y-axis of the representation. For example, the type of an audio signal (time to amplitude representation) is "t:a", whereas the type of a spectrum (frequency to amplitude) is "f:a".

Vectors are special cases of functions, associating an index to a value. Because vectors are very frequent, we introduce the shortcut symbol "V + data type" to denote a vector. For instance, a list of time onsets in an audio signal is notated "Vt", or the type of a signal split into frames is "Vt:a".

Typing Rules. The output types of the operators are computed dynamically in a bottom-up fashion and recursively according to specific typing rules depending on the type of the input data. For instance, in the case of vectors, the typing rule is: "Type (F(Vx)) = V Type(F(x))".

For non-vector arguments, each operator defines a specific typing rule. For instance:

- the output type of Abs is the type of its input: Type (Abs(arg)) = Type(arg)
- the FFT operator multiplies the x-axis dimension of its input by -1: Type (FFT (a:b)) = a-1:b, thus transforms "t:a" into "f:a", and reversely "f:a" into "t:a"
- splitting some data introduces a vector of the same data type: Type (Split (x)) = V Type(x),
- and so forth.

This typing system is more complex than the usual typing systems used routinely in GP, but has the interest of being able to retain the respective physical dimensions of the inputs and outputs values of functions. For instance, given an input signal S, the following complex (but realistic) function has the following type: Type (Min(Max(Sqrt(Split (FFT(Split (SIGNAL, 3, 100)), 2, 100)))) = "a" Furthermore, this type can be inferred automatically by the typing system using the typing rules.

Generic Operators Patterns. Besides ensuring the consistency of operators, the typing system can be used proactively, to specify function spaces. The idea is to specify complex expressions partially, using the type system. More precisely, a we introduce the notion of "generic operator", denoting a set of operator(s) whose output types (and also possible arguments) are specified syntactically. Three different generic operators (notated "*", "!", and "?") have been introduced, with different functionalities:

- "?_T" stands for 1 operator whose output type is "T"
- "*_T" stands for a composite operator for which the output types of its constituents are all "T"
- "!_T" stands for a composite operator for which only the final output type is "T"

These generic operators allow to write functions patterns, that stand for any function satisfying a given signal processing method. For instance, the pattern: "?_a (!_Va (Split (*_t:a (SIGNAL))))" stands for:

- "Apply some signal transformations in the temporal domain" (*_t:a)
- "Split the resulting signal into frames" (Split)
- "Find a vector of characteristic values - 1 for each frame" (!_Va)
- "Find one operation to find one relevant characteristic value for the entire signal" (?_a)

This pattern represents in fact the general extraction scheme presented in 1.4. It can be instantiated as:

- Sum_a (Square_Va (Mean_Va (Split_Vt:a (HpFilter_t:a (SIGNAL_t:a, 1000Hz), 100)))), or
- Log10_a (Variance_a (NPeaks_Va (Split_Vt:a (Autocorrelation_t:a (SIGNAL_t:a), 100), 10)))

Because patterns denote function sets, they can be specified in the EDS algorithm to guide the search of functions, to force the system to create only functions satisfying a given pattern, as seen below.

2.3 EDS Algorithm

The global architecture of the EDS system consists in 2 steps (see Fig. 4):

- Learning of relevant features using a genetic search algorithm,
- Validation of features and synthesis of a descriptor extractor using features combination.

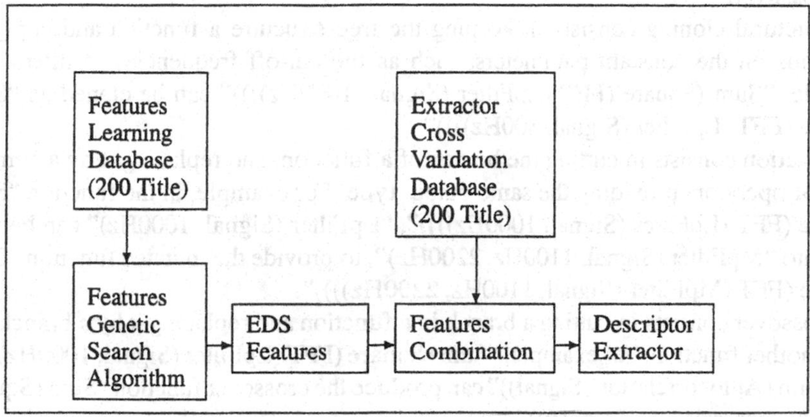

Fig. 4. EDS Global Architecture

EDS search algorithm is based on genetic programming, i.e. the application of genetic search to the world of functions, as introduced by Goldberg ([12]). More precisely, the algorithm works as follows, given:

- A descriptor D for which we seek an extractor, and its type (currently either "Boolean" or "Float"),
- A database DB containing audio signals,
- A result database containing the result of the perceptive test for the descriptor D for each signal in DB.

Global Algorithm. The algorithm proceeds as follows:

- Build the first Population P0, by computing N random signal processing functions (compositions of operators), whose output type is compatible with the type of D.
- Begin Loop:
  - Computation of the functions for each audio signal in DB,
  - Computation of the fitness of each function, for instance the correlation between its values on DB and the associated perceptive values,
  - if the (fitness >= threshold) or (max number of iterations reached), STOP and RETURN the best functions,
  - Selection of the most correlated functions, crossover and mutation, to produce a new population P_{i+1}
  - Simplification of the population P_{i+1} with rewriting rules
  - Return to Begin Loop

Creation of Populations. To create initial populations, a random function generator creates functions according to a given pattern (as described in 2.2). The generator works bottom-up, starting with the audio signal as the main argument, and finding successively operators that accept the current operator as input. New populations are then computed by applying genetic operations to the most relevant functions of the current population. The operators used are structural cloning (constants variations), mutation, and crossover.

Structural cloning consists in keeping the tree structure a function and applying variations on the constant parameters, such as the cut-off frequencies of filters. For example, "Sum (Square (FFT (LpFilter (Signal, 1000Hz))))" can be cloned as "Sum (Square (FFT (LpFilter (Signal, 800Hz))))".

Mutation consists in cutting the branch of a function, and replacing it by a composition of operators providing the same output type. For example, in the function "Sum (Square (FFT (LpFilter (Signal, 1000Hz))))", "LpFilter (Signal, 1000Hz)" can be mutated into "MpFilter (Signal, 1100Hz, 2200Hz)", to provide the mutated function "Sum (Square (FFT (MpFilter (Signal, 1100Hz, 2200Hz))))".

Crossover consists in cutting a branch in a function and replacing it by a branch cut from another function. For example, "Sum (Square (FFT (LpFilter (Signal, 1000Hz))))" and "Sum (Autocorrelation (Signal))" can produce the crossover function "Sum (Square (FFT (Autocorrelation (Signal))))".

Eventually, to ensure diversity , new populations are completed with a set of new random functions.

2.4 Heuristics

Heuristics are vital ingredients to guide the search and a central point in the design of EDS. They represent the know-how of signal processing experts, about functions seen a priori, i.e. before their evaluation. The interest of heuristics is that they both favor a priori interesting functions, and rule out obviously non-interesting ones.

A heuristic in EDS associates a score between 0 and 10 to a potential composition of operators. These scores are used by EDS to select candidates at all the function creation stages (random, mutation, cloning and crossover). Here are some examples of important heuristics:

- To control the structure of the functions: "HpFilter (Signal, Branch) ⇒ SCORE = Max (0, 5 - Size(Branch))". This heuristic limits the complexity of computation of arguments such as filters cut-off frequencies.
- To avoid bad combination of filters: "HpFilter (HpFilter ⇒ SCORE = 1", "Mp (Hp⇒ 3", "Lp (Hp ⇒ 5".
- To range constant values: "Enveloppe (x, < 50 frames) ⇒ SCORE = 1"; "HpFilter (x, < 100Hz) ⇒ 1", etc...
- To avoid usually useless operations: "X (X (X ⇒ SCORE = 2" (too many repetitions of operators), etc...

2.5 Rewriting Rules

Rewriting rules are applied to simplify functions before their evaluation, using a fixed point mechanism until to obtain a normal form. Unlike heuristics, they are not used by the genetic algorithm to favor combinations, but:

- Avoid computing several times the same function with different but equivalent forms. For example: "Correlation (x, x) ⟹ Autocorrelation (x)", or "HpFilter (HpFilter (x,a), b) ⟹ HpFilter (x, max(a, b))".
- Reduce the computation cost. For Example: Perseval equality "Mean(Fft(Square (x))) ⇒ Sum(Square (x))" avoids to compute the "Fft" of a signal.

2.6 Caching

Finally, to speed up the computation of functions, a caching mechanism is introduced, so that any costly function is computed once, and reused when possible. Every time a new function is computed, all the intermediate results are stored on separate files. For instance: "Max (Envelope (Fft (x), 100)" will store "x", "100", "Fft(x)", "Envelope (Fft (x), 100)", and "Max (Envelope (Fft (x), 100)" for each tested title.

The caching technique consists in keeping in memory the most useful results, depending on:

- their computation time: results that require a long computation time are kept in memory,
- their utility: results that are used frequently are kept,
- their size: the allowable memory being limited, priority is given to small size results.

3 Results

We present here the results of the 2 steps of EDS:

- Features computation (learning results): the correlation of the best functions found by the system evaluates how our genetic search algorithm is able to build relevant functions regarding a given data set. Correlations are computed on the whole features learning database.
- Descriptor extraction (test results): final model error (regression) or final recognition rate (classification) for one or a combination of the N most relevant functions. Performances and errors are evaluated using cross-validation on an independent test database, so the results are given with an uncertainty that corresponds to the cross-validation variations. For each of these descriptor, we compare the results obtained by the traditional LLD method, the EDS method, and a combination of both.

3.1 Sinus + Colored Noise (Basic Problem)

The problem consists in detecting a sinus between 10 and 1000Hz mixed with a strong colored noise between 1000-2000Hz (see 1.4) (Of course, the problem must be solved only by looking at the test database).

Best Features and Extractors. As they focus on the most predominant characteristic of the signal (the noise), the LLDs have poor results in detecting the sinus. The best LLD, Spectral Flatness, has a correlation of 0.63 with the sinus frequencies.

EDS finds not only a very good extractor, but also an extractor that makes sense, given the problem at hand. More precisely, EDS focuses after 10 populations around the function "MaxPos (FFT (LpFilter (Signal, fc Hz)))", with different values of fc. Values between 50 and 700 Hz, that most efficiently remove the colored noise (with a Butterworth filter), provide a correlation of 0.99. The correlation does not reach 1 because of the uncertainty near 1000Hz. For the Spectral Flatness, the mean prediction error is 226Hz, whereas it is 10Hz for the best EDS function.

3.2 Perceived Intensity (Regression Problem)

The problem consists in providing a model of the subjective energy of musical extracts, based on the results of perceptive tests (see 1.3). For comparison, note that a random feature has typically a correlation of 0.03, and its best combination provides a model error of 21% (the extraction function is a constant value, that is the mean value of the energies of all the titles in the database).

Best Features and Extractors. The best LLD found has a correlation of 0.53 with the perceptive values, and provides a model error 16.9%+-1.5%. The best EDS feature has a correlation of 0.68 and provides a model error 14.5%+-1.8%. The best combination of LLD provides a mean model error of 12.1%+-1.9%. Adding the best EDS features decreases the mean model error to 11.4%+-1.9%.

Table 2. Model Errors for the Subjective Energy Problem

method	Random	Best LLD LLD	Best LLDs Combination	Best EDS	Combination LLDs + EDSs
Perceived Energy (Model Error)	21%	16.9%	12.1%	14.5%	11.3%

This shows a clear improvement of EDS over the standard approach on a typical high-level extraction problem.

3.3 Presence of Voice (Classification Problem)

The problem consists in providing an extractor that detects the presence of singing voice (see 1.3). Note that a random feature has typically a correlation of 0.05, and its best combination provides a recognition rate of 50% (the extraction function is a constant value, that assigns the same result to all the titles in the database).

Best Features and Extractors. The best LLD has a correlation of 0.28 with the perceptive values, and has a recognition rate of 61.0%+-10.2%. The best EDS feature has a correlation of 0.54, and provides a recognition rate of 73.5%+-9.4%. The best combination of LLD provides a recognition rate of 63.0%+-11.4%. Adding the best EDS features increases the recognition rate to 84.0%+-7.7%.

4 Conclusion

We have introduced a new approach for designing high level audio feature extractors, based on genetic programming. The proposed system, EDS, uses for the moment a limited palette of signal processing functions. However, EDS produces results that are as

Table 3. Recognition Rates for the Presence of Singing Voice Problem.

method	Random	Best LLD LLD	Best LLDs Combination	Best EDS	Combination LLDs + EDSs
Presence of (Singing Voice) (recognition Rate)	50%	61%	63%	73.5%	84.5%

good or best as standard, manual approaches in high level descriptor extraction. Substantial increase in performance should be obtained by extending the palette of signal operators to more refined operators. New heuristics will also be found by analyzing the application of EDS to other high level descriptor problems, such as the distinction between "live" and studio recording, the discrimination between simple and generic genres (such as military music, music for children, etc.), the estimation of the danceability, etc. Also, better fitness method can be used, including in particular a fully-fledged learning mechanism to match optimally the outputs of the functions to perceptive tests.

References

1. Eric D. Scheirer. Tempo and beat analysis of acoustic musical signals. J. Acoust. Soc. Am. (JASA) 103:1 (Jan 1998), pp 588-601.
2. Eric D. Scheirer, and Malcolm Slaney. Construction and evaluation of a robust multifeature speech/music discriminator. Proc ICASSP '97, pp. 1331-1334.
3. P. Herrera, A. Yeterian, F. Gouyon. Automatic classification of drum sounds: a comparison of feature selection methods and classification techniques. Proceedings of 2nd International Conference on Music and Artificial Intelligence, Edinburgh, Scotland, 2002.
4. Geoffroy Peeters, Xavier Rodet. Automatically selecting signal descriptors for sound classification. Proceedings of the 2002 ICMC, Goteborg (Sweden), September 2002.
5. Perfecto Herrera, Xavier Serra, Geoffroy Peeters. Audio descriptors and descriptors schemes in the context of MPEG-7. Proceedings of the 1999 ICMC, Beijing, China, October 1999.
6. A.L. Berenzweig, Dan P. W. Ellis. Locating singing voice segments within music signals. IEEE workshop on applications of signal processing to acoustics and audio (WASPAA01), Mohonk NY, October 2001.
7. Wu Chou and Liang Gu, "Robust Singing Detection in Speech/Music Discriminator Design," International Conference on Acoustics, Speech, and Signal Processing (ICASSP 2001), pp.865-868, Salt Lake City, Utah, USA, May 2001
8. JJ Aucouturier, François Pachet. Music similarity measures: what's the use ? In proceedings of the 3rd international symposium on music information retrieval (ISMIR02), Paris, October 2002.
9. George Tzanetakis, Georg Essl, Perry Cook. Automatic musical genre classification of audio signals. Proceedings of 2nd International Symposium on Music Information Retrieval, pp 205–210, Bloomington, IN, USA, October 2001.
10. John R. Koza. Genetic Programming: on the programming of computers by means of natural selection. Cambridge, MA: The MIT Press.
11. David J Montana. Strongly typed genetic programming. In Evolutionary Computation 3-2, 1995, pp 199-230.
12. David E. Goldberg. Genetic algorithms in search, optimization and machine learning. Addison-Wesley Pub. Co. 1989. ISBN: 0201157675.

'Ambiguous Live' – Exploring Collaborative, Dynamic Control of MIDI Sequencers

David Eyers

Computer Laboratory
University of Cambridge
David.Eyers@cl.cam.ac.uk

Abstract. Computer sequencers generally control other digital musical instruments. This paper discusses our particular interest in playing the sequencers themselves as digital musical instruments. We present our ensemble's approach to live sequencer-based performance, supporting multiple musicians coordinating activities through one central control system. We then review a number of promising software technologies currently available for the purpose of such live sequencer-based performances, and finally discuss our consequent future performance goals.

1 Introduction

People have gathered together to enjoy musical performances for thousands of years. Many of these performances have included groups of musicians who, through improvisation or interpretation, cooperate to resonate their ideas and musical intuition. Modern day orchestras and bands combine the talents of their constituent musicians to form a dynamic and complex higher-order performance entity.

How does this compare with the music flourishing throughout the night-clubs of the world today, however? Many have no respect for this 'dance' music, due to its prefabricated nature. However, we feel that proliferation of digital musical technology is an opportunity rather than an affront to music performance (or it is at least both!).

Whilst hardly a house-hold name, 'Ambiguous Live', was a duo formed in part by the author for the sake of investigating the ease with which we could design software, supported by a hardware rig of synthesisers, to enable live, largely improvised electronic music performance. Our approach involved the high-level control of a computer sequencer, centrally coordinating the actions of multiple musicians. In the group's brief lifetime, we performed in a number of university-oriented Sydney nightclubs, and were interviewed regarding our live operating environment in the street magazine '3DWorld' under its 'localz' section in March 2001 [1]. Unfortunately this group had to suspend activities when the author left Australia to pursue postgraduate education in the UK.

This paper is organised as follows. We provide some background discussion relating to electronic music performance in section 2. We discuss our performance approach in section 3. A number of technologies which support our goals to a greater or lesser extent are reviewed in section 4. Section 5 presents our plans and desires for future performance environments. Finally we conclude the paper in section 7.

U.K. Wiil (Ed.): CMMR 2003, LNCS 2771, pp. 54–63, 2004.

2 Background

Disc Jockeys (DJs) originally had the job of setting up the media to be played in sequence for radio stations. They may have provided voice interludes between the songs, and indeed may have had a greater or lesser degree of control in the actual sequence of music played. Although possibly very musically knowledgeable, they were not musicians in any performance sense.

Nightclubs also need someone to keep music playing when there are no live bands, so the role name was transfered to this environment, although the job being performed varies greatly. Some DJs merely choose an order of tracks to play, reading the mood of the crowd but nothing more than that in a performance sense. These DJs are sometimes bestowed undue credit; the audience members are actually enjoying the work of the composers rather than the DJs, but the DJs provide a face to which to direct appreciation. However, most club DJs interact more directly with their performance machinery, controlling the speed and looping of a number of sound sources – they simultaneously perform audio mixing and compositing. Compact Disc (CD) players (or possibly computer-based multi-track MP3 players) may be used by comparatively unskilled DJs, although most serious DJs use traditional phonograph turntables. Undoubtedly good turntable control is a skilled role, and there is ample space for artistic creativity. However, the musical building blocks with which such DJs perform are still very coarse in musical granularity (at the level of whole tracks or phrases).

The most sophisticated DJs in fact render their own musical material, and indeed are the artists responsible for making the material other DJs mix and play back to their audiences. However, even these DJs have only limited abilities to create live music performance. Some have taken to including drum machines, or other simple sequencers into their performance rigs – we feel that is an initial step to acknowledging the relevance of sequencer performance discussed in this paper.

3 Ambiguous Live

Our performance goal was to be able to create synthesised music via live collaboration between performers, with a heavy emphasis on improvisation. Limitations in our software sequencer meant that we did need to prerecord a number of musical patterns, however for most monophonic parts we divorced the rhythmic control from the tonal control. Essentially we programmed the synthesiser to convolve separate rhythmic and melodic patterns into MIDI events, in effect creating a simple Control Voltage (a MIDI precursor) to MIDI converter.

Our equipment rig grew in between each of our performances. In our most recent performance, our inventory included three MIDI-controller keyboards (in the piano sense), two general purpose MIDI controller knob banks, two samplers, two analog modelling synthesisers (one monophonic, and one 16-channel multitimbral unit), a multitimbral digital synthesiser, a drum machine, four effects processors, two mixers, a Sony VAIO laptop, and a desk fan.

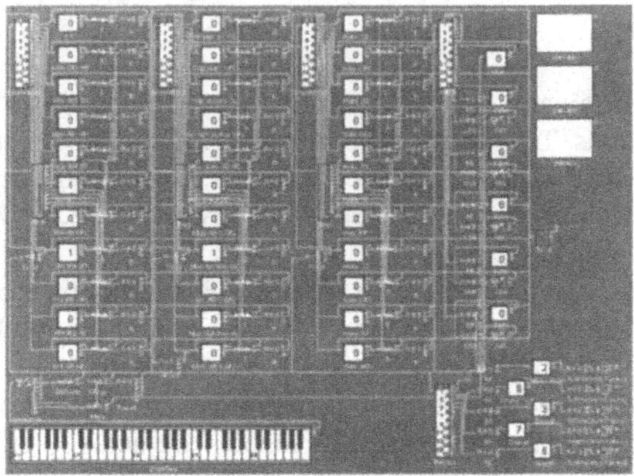

Fig. 1. The 'Ambiguous Live' rhythm control environment screen-shot

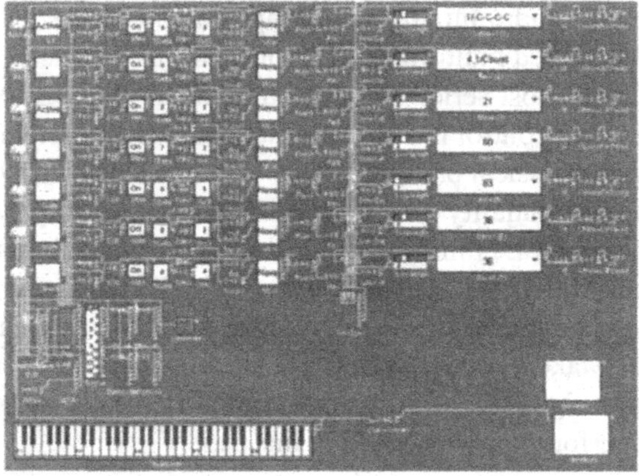

Fig. 2. The 'Ambiguous Live' stream control environment screen-shot

3.1 Music Control Structure

Our top-level control architecture consisted of two main regions. One whole control area in our environment was devoted to rhythmic parts (drums and playback of prerecorded samples); a screen-shot is shown in Figure 1.

The other main control area related to numerous 'streams'. We used this term so as not to confuse streams with MIDI channels; any stream could output to any number of MIDI channels on any of our devices. Essentially each stream was responsible for a section of our music. A screen-shot of the stream environment is shown in Figure 2. We generally used the following streams:

Sub-bass. Used for very low bass used predominantly to keep sub-woofer speakers occupied, generally for the sake of shaking the performance venue rather than making particularly pitched contributions.

Bass. Provide audible bass-lines providing low-pitch texture and rhythm.

Noise. This stream was employed to play signature samples or noises to identify each particular part of a musical set.

Pads. Used for fairly soft-timbred chords providing a harmonic foundation.

Arps. Arpeggios are often used to produce high-pitch texture and rhythm.

Lead. Synthesiser lines which stand out well in front of the rest of the music. Usually monophonic.

Note that while we were using a sequencer which provided a piano-roll style view, the actual music output did not come directly from the time-line. Some control parts from the time-line were used to influence pattern playback however. For example patterns of different length aligned themselves on specific bar numbers, ensuring that changing from an 8 bar to a 1 bar and back to an 8 bar pattern would not cause the second 8 bar pattern to be played out of alignment with other 8 bar patterns.

3.2 Music Control Inputs

Another experimental aspect of our performances was that the centralised sequencer permitted both performers to collaborate, interacting across all the controls. There were several ways in which we could exert control over our music sequencing environment:

Via laptop. The mouse and keyboard of the laptop running our sequencer could be used to adjust parameters and override controls. We were only using one physical display screen, so needed to use computer keyboard hot-keys to change which virtual screen of data within the sequencer we were examining or manipulating.

Via direct synthesiser controls. All of our audio synthesis equipment was present at each performance, so we were able to directly manipulate each device's controls.

MIDI dial boxes. We used two banks of assignable MIDI dial boxes to permit us to control parameters on the equipment. We programmed the sequencer to provide a number of default pages of knob assignments, but included the capacity for knobs to be assigned by the laptop during performance.

MIDI keyboards. In the relative dark, we found it easiest to use three MIDI piano keyboards between us for both melodic performance, and for remote control of the sequencer. On the melodic keyboards, upper octaves played notes in various lead streams. The lowest octave was used for real-time transposition of streams currently set to have non-fixed root notes. Thus if the sequencer was playing some particular bass-line, and an introduced chord progression required its transposition to fit harmonically, it was sufficient for a performer to tap an appropriate lower octave key once to set a new root note. The middle octaves of the keyboard were reserved for muting, routing and assignment control. The mute controls allowed stream output to be switched on and off easily. The routing controls allowed each stream to be moved dynamically to other synthesisers, and finally the assignment controls allowed any of our prerecorded patterns to be assigned to any of the streams.

Note that to compress so many different functions into limited numbers of keys, chording was required (and we add works far better for keyboard-players on musical keyboards than on computer keyboards!). So, for example, three stream select notes could be played simultaneously with the mute key to affect them all together.

Our sequencer environment was created within Emagic's Logic Audio [4] software sequencer. The most remarkable feature of this program is its graphical MIDI programming environment. The next section discusses why we believe graphical environments are so well suited to sequencer programming and our type of application development.

3.3 Graphical Programming

Unlike traditional procedural programming in which one considers the computational steps sequentially, music performance, and thus sequencer programming, tends to be much more event driven. In effect, the performance environment can be thought of as a web of interconnected nodes, each of which responds to some sort of message or set of messages sent to it on its inputs, by itself generating some number of modified events on its outputs. Usually there is only very simple state held within each of these nodes. For MIDI sequencer programming, the environment is even more predictable, since the structure of the MIDI messages themselves are very well defined.

Emagic's Logic Audio package provides for the programming of MIDI 'environments', which consist of a number of types of nodes connected together visually by 'wires' which indicate where events will flow. The language is reasonably expressive; transformer nodes can themselves be modified by 'meta-events'; events intended only for the internal control of the sequencer.

Crucial to our programming were the Logic 'touch-tracks' node type, which allowed us to trigger the hundreds of simultaneous sequenced parts we needed as event sources under the control of the environment itself (visible on the right hand extreme of each stream path in Figure 2). In fact, the conventional sequencer time-line merely contained some muted emergency parts, and some parts providing control triggers into the environment.

Apart from some of the limits we reached in using Logic (discussed in section 5), we generally feel that graphical programming is particularly suited to live sequencer design and operation. By using two-dimensional spatial layout, a great deal of information can be compressed into each screen. Also, there is little distinction between design time and control-time. Graphical components can be controlled during performances in the same manner in which they are built up during sequencer environment programming.

4 Alternative Tools for Developing Live Performance Environments

We feel our performances merely scratched the surface of this fascinating potential area of live music. Due to availability and time limitations, we used the existing Logic sequencer as discussed above. There are a number of other promising current technologies which might be employed to perform music in a similar manner, some of the ones we experimented with before choosing Logic are described in this section.

4.1 Collaborative Music Environments

One of the most wide-spread collaborative music environments offered recently has been the Rocket Network [10]. Unfortunately the Rocket Network service providers closed down their service in early 2003. The RocketControl™ software hooks into commercial sequencers to allow composers to conveniently exchange musical data, both in MIDI and audio forms. Some sequencer versions have been released which cater directly to this function, such as Logic Rocket, and Cubase InWired. The Rocket Network also provided a centralised storage facility for media.

There are numerous other musical collaboration sites on the Internet, but they generally offer a loose coupling, basically implementing some form of discussion board or Internet data storage site through which files can be passed back and forth between artists. However web browsers provide a very general interface to such services and suffer from not being able to cater specifically for the demands of music media.

Note that all of the above forms of musical collaboration environment were fundamentally unsuitable for our performances because they introduce a delay in the exchange of data and control. Whilst they provide great opportunities for globally distributed artists to work together with relative ease, they do little to help two performers standing next to each other to lever their mutual control of musical equipment.

4.2 Software Studios

Due to the recent surges in computer multimedia processing speed, it is now practical for a computer to do sufficient software synthesis to have prompted a number of companies to develop virtual studio packages.

Reason. Propellerhead Software's 'Reason' [9] is an excellent example of a virtual studio package. It provides the user an ability to create as many virtual synthesisers as any given computer is capable of producing faster than its audio streaming speed. Synthesisers can be linked together via a virtual patching system implemented within the software (and again, using a graphical view of signal flow).

For our purposes this software did not provide enough flexibility in the MIDI programming domain however. It might provide an excellent low-cost substitute for many types of MIDI synthesiser, but would really just perform as a target instrument in the rigs we have built.

Orion. Synapse Audio Software's 'Orion' [11] also deserves a mention. It provides the same excellent core of software synthesis and effects limited only by the speed of the computer on which it runs. However, the ability to experiment with custom MIDI and audio control and routing is more limited than that of Reason. So again, although the sound quality produced was truly impressive, we could not use this a package for any purpose other than as a very low-cost target MIDI instrument.

Cubase VST. Before designing our final performance environment in Emagic's Logic sequencer as we described above, we also experimented with Steinberg's Cubase VST (Virtual Studio Technology) software [12].

Whilst Cubase VST is an excellent package, which we found easy to use and which features good audio and MIDI integration, it unfortunately lacked programmability to the degree we required for our live sequencer performance. It has some controls which play patterns rather than just notes when keys are pressed, but without the ability to re-program MIDI event routing and transformations from within the sequencer environment itself, we were unable to live control it sufficiently for our needs.

4.3 Off-Line Programming Environments

We found it particularly convenient to have little distinction between programming and performance contexts, since development was an on-going process. However, if the design of your desired sequencer performance machines are fairly static, a number of tools exist for easing the programming of such applications.

One notable programming environment is C-Sound [3]. Audio and MIDI based performance systems can be designed in its programming language, and then compiled to make operating machines. As with most programming, the compilation step potentially provides a significant performance boost, but at the cost of preventing dynamic adjustment of the application's software.

Naturally general-purpose programming languages could also be employed to develop performance machines. This was not a viable option for our performances; our music machine design was constantly being extended, and our development time was severely limited.

In terms of reducing the programming load (at a speed penalty), the Sun's Java language [6] provides library functions to communicate with MIDI devices. Thus a significant amount of the machine-specific programming can be avoided and the programmer left to focus on the application. It also has the advantage of portability to other machines. As a real-time environment, however, even if we had had the time required to develop such custom software, we were sceptical that Java's timing would have been fast or reliable enough for our performance needs. Even so, JavaBeans provide an interesting potential component framework in which to move the flexibility of Java code modules into a live, graphical programming environment. Since it is already embedded in an event-driven framework, it is described in the next section.

4.4 Live Programming Environments

We noted a number of other potential live, graphical music-oriented programming environments described on the Internet. Cycling'89's Max and Max/MSP [13] products look extremely promising, but we only had Windows platform computers, for which they have not yet released their software.

The Reaktor software package from Native Instruments [5] is another live, graphical musical programming environment, but tends to be more focused in the audio synthesis domain, rather than the manipulation of MIDI events. Whilst a very flexible and powerful transformation tool, Reaktor's use for us would have been complicated by its lack of the musical editing views provided by software sequencers such as Cubase or Logic.

We have been interested to see many large software vendors recently proposing component-based development architectures. Although most developers have not yet fully embraced such programming ideals, we feel that the domain of MIDI event processing and performance is particularly suitable to component-based design; indeed the environments in Logic employ its internal component-based toolkit.

As mentioned above, we are concerned that the real-time properties (for example its garbage collection strategy) of Java are not particularly desirable for performance. Such concerns aside, Sun's JavaBeans Component architecture for Java [7] may provide a useful framework in which to develop dynamic MIDI machines such as our performance environment. At the moment no MIDI-related components are listed in Sun's

JavaBeans on-line directory, but their Bean Builder application may indeed eventually surpass the functionality of the Logic environment through extensibility. We feel the most likely problems with JavaBeans design will be the Bean Builder interface being polluted by unnecessary component details, or possibly that the change in and out of the design mode will be too disruptive to allow the MIDI event handling to continue while the application is being extended or modified.

5 Extensions

This section describes a number of our future plans, based on where we were not able to to work around the restrictions imposed on us.

5.1 Live Recording Elements

The biggest limitation we faced was that we were not able to record passages for looping back during our set. For example, we might improvise an eight bar chord pattern via a sequence of piano key presses in an octave we programmed to set the transposition root note of certain streams. Ideally, the software environment would support elements with memory which would maintain and repeat this eight bar pattern until further input was received to override it. Our performances only had the ability to playback prerecorded patterns, although as mentioned above these patterns could span granularity from a simple rhythm up to a complete musical phrase.

The Logic environment would potentially permit the programming of a very simple step-wise live-recording sequencer, but the amount of extra environment design work required would be massive, and rather redundant given the touch-tracks node type is already offering a read-only version of the service we want, plus the code to handle recording is already implemented to run the conventional Logic sequencer time-line.

Note that such an element would not merely provide the ability to live-record and loop musical patterns; more importantly the control stream events which affect the sequencer environment itself would be able to be recorded. Uses might include recording a list of chord sequences which can be mapped and unmapped to any stream, as mentioned above, or indeed changing the timbre or other parameters of a particular synthesiser in a periodic manner.

5.2 Unified Design and Performance Contexts

Generally, when we had started a particular performance set, we would not make any changes to the design of our environment. Whilst one would not normally expect to make modifications, we found that certain changes within the Logic environment would unsettle other events passing through the system. Ideally, like many of the other live music or audio programming products discussed above, there would be no distinction between design and performance environments. Placing new objects within the sequencer programming should not affect the current live function of the sequencer at all.

A further related area of interest is in optimisation. Because we only had limited internal control over the MIDI message patterns being played by the touch-tracks elements, we sometimes needed to carry unnecessary information into the sequencer. For example, one of our touch-tracks elements might play a particular stream in all 16 MIDI channels, when in fact all but one of these channels would be filtered out before reaching the sequencer's MIDI output device. An interesting challenge would be to design the sequencer so that it could 'push-back' such filter events as early as possible within the sequencer programming, thus reducing unnecessary MIDI message handling. Generally due to the cheapness of CPU cycles, this area has not been explored. In computer networks, however, packet handling can have very high costs. We are currently working with a research group in publish/subscribe middleware (see [8]) with a view to possibly employing such technologies as composite event detection and subscription spanning trees for live, collaborative performance arts.

5.3 Integration of Other Performance Media

Our future plans include performances with greater integration of music, lighting, and video media (where present). Most DJs have their music performances enhanced by lighting and/or video displays. In the same manner as for our discussion of DJs above, however, there is often only a token connection between these different performance areas.

Instead of having three people controlling the music, the lighting and video projections separately, we are interested in programming performance environments where, in the manner of our collaborative performances, these three separate people together control one sequencer, which then itself splits back out to control separate synthesiser, lighting and video equipment. This integration is intended to lift the lighting operator, say, from the level of control required to get lighting effects basically in time with the music, to deciding at a much higher, and more artistic level, how different sections of music will relate to the activity of the lighting fixtures. Naturally override controls would still be accessible at a device level for emergencies. Normally, however, the timing aspects would be left to the centralised event sequencer, since it already needs to understand and process the temporal context with respect to notes, beats, bars and phrases for the music output.

6 Our Live Sequencer-Based Collaboration Feature Checklist

Given that we were not able to satisfy all our needs with the software available, we felt we should indicate what our requirements would be:

Live Performance Speed. Many packages induce delays when incorporating collaborative input. We require that delays are imperceivable to live performers.

Unified Design and Performance Control. It is important that the musical collaboration environment make few distinctions between the performance and design phases. A performer should be able to modify the environment while another performer is playing it with little or no disruption.

Graphical Event Flow Views. We found it highly intuitive to program our performance machines in a graphical environment where events could be traced through an interconnected network of processing nodes.

Music Specific Interfaces. Many component programming environments do not yet provide convenient support for management of musical patterns (or score), audio, and the consequent temporal synchronisation and arrangement of these sequence blocks. Such interfaces are necessary to facilitate rapid composition.

Interface Design and Modularisation. It is highly useful to support encapsulation of processing networks into higher-level modules. These modules need to have user-programmable parameter interfaces.

7 Conclusion

This paper discussed modern popular electronic music performance, and suggested that developing live control environments around music sequencers is a promising new area of ensemble-based musicianship. We presented an overview of the performance environment used by the small Sydney group 'Ambiguous Live' formed in part by the author. A number of alternative current technologies potentially supporting the development of such environments are discussed. Finally, we propose a number of future directions in which we want to develop both our performances, and research into the music technology that will power them.

References

1. 3DWorld. Blurred beats - ambiguous live. `http://www.threedworld.com.au/`, March 2001.
2. MIDI Manufacturers Association. *The Complete MIDI 1.0 Detailed Specification*. MIDI Manufacturers Association, 1996.
3. Richard Charles Boulanger, editor. *The Csound Book: Perspectives in Software Synthesis, Sound Design, Signal Processing,and Programming*. MIT Press, 2000.
4. Emagic. Logic audio. `http://www.emagic.de/products/ls/ls/index.php`.
5. Native Instruments. Reaktor. `http://www.nativeinstruments.de/index.php?reaktor_us`.
6. Bill Joy et al., editors. *Java™ Language Specification*. Addison-Wesley, second edition, June 2000.
7. Sun Microsystems. JavaBeans™ API specification. `http://java.sun.com/products/javabeans/docs/beans.101.pdf`.
8. Peter R. Pietzuch and Jean M. Bacon. Hermes: A Distributed Event-Based Middleware Architecture. Proceedings of the 1st International Workshop on Distributed Event-Based Systems (DEBS'02), July 2002.
9. Propellerhead Software. Reason. `http://www.propellerheads.se/products/reason/`.
10. Rocket Network Inc. The RocketControl™ online audio collaboration tool. `http://www.rocketnetwork.com/`.
11. Synapse Audio Software. Orion. `http://www.synapse-audio.com/products.php`.
12. Steinberg. Cubase. `http://cubase.net/`.
13. Todd Winkler. *Composing Interactive Music: Techniques and Ideas Using Max*. MIT Press, 2001.

Characterization of Musical Performance Using Physical Sound Synthesis Models

Philippe Guillemain and Thierry Voinier

CNRS-Laboratoire de Mécanique et d'Acoustique
31, chemin Joseph Aiguier. 13402, Marseille cedex 20, France
guillem@lma.cnrs-mrs.fr

Abstract. Sound synthesis can be considered as a tool to characterize a musical performance together with the performer himself. Indeed, the association of real-time synthesis algorithms based on an accurate description of the physical behavior of the instrument and its control parameters, and gestures capture devices the output parameters of which can be recorded simultaneously with the synthesised sound, can be considered as a "spying chain" allowing the study of a musical performance. In this paper, we present a clarinet synthesis model the parameters and controls of which are fully and explicitely related to the physics, and we use this model as a starting point to make a link between the playing and the sound.

1 Introduction

Self-oscillating musical instruments, such as woodwinds, possess a very rich performance potential, since several playing parameters are under continuous control of the musician. This richness is an advantage in terms of the sound palette, but is also a difficulty for the scientist willing at qualifying a performance. Indeed, though the sound (*i.e* the effect) resulting of the interaction between the player and the instrument contains all the information required for a subjective qualification of a performance, objective measurements of playing parameters (*i.e* the source) provide complementary informations that can help understanding the musical intention. Unfortunately, recovering physical playing parameters from the analysis of a sound sequence is still an open problem, even in the case of "simply driven" instruments such as the piano. For this instrument, in order to circumvent this problem, electronic capture devices combined with the real instrument have been designed, leading for example to the famous Disklavier from Yamaha, used *e.g* to study the influence of the room acoustics on the pianist playing [Bol]. Capturing gestures in the same way on real woodwind instruments requires a more complex instrumentation that may changes the feeling of play. For example, the blowing pressure is obtained by measurement of the static pressure within the mouth of the player, requiring invading techniques.

In this framework, digital techniques for real-time simulation of musical instruments can be useful. The use of sound synthesis models based on the physics

U.K. Wiil (Ed.): CMMR 2003, LNCS 2771, pp. 64–73, 2004.

of the instrument becomes necessary in this case since it allows a behavior of the digital instrument faithfull with respect to the real one. In the same way, digital controllers offering the ability of accurately piloting the synthesis model and giving a natural feeling of play to the musician, have to be used.

Such an association of a suitable controller and a synthesis method, constituting an artificial instrument, can be considered as a tool for performance and performer classification. Indeed, capturing simultaneously the continuous action of the player on the controller and the sound produced by the synthesis model, is a way to collect datas representative of the conscious and unconscious feedback operated by the player in order to produce the desired sound. Such playing datas can then be used as input for statistical processing, in the aim of classification of performances and signature of players. In other words, similarily to the artificial mouthes developed to study the physical behavior of real woodwind instruments under calibrated playing conditions, the main statement of this paper is the opposite, but complementary situation consisting in using an artificial instrument in order to study the behavior of a real player.

To achieve these goals, a real-time synthesis model based on physical models is needed, in order to make an explicit link beetween the geometrical and physical playing parameters, and the synthesis parameters. In the clarinet case, the physical behavior of the instrument is commonly decribed by a set of three equations: a linear equation linking the acoustic pressure and flow at the mouthpiece; a linear equation linking the reed displacement and the acoustic pressure; a nonlinear equation linking the flow, the acoustic pressure and the reed displacement. In this case, the calculation of the self-oscillations has first been proposed by Schumacher [Sch], using a time domain discretization of the equations. Such a work that does not attempt to a real-time implementation has been since widely developped [Gaz], [Duc], [Ker1]. In order to perform the calculations in real-time, several methods yielding a time domain formulation of the waves in the resonator have been developed. We can mention the so-called digital waveguide method (see for example Smith [Smi] or Välimäki [Väl]). Digital Wave filters have also been applied in this context [Wal]. Like the Schumacher method, these methods consider the incoming and outgoing waves within each bore section and their scattering at the interface between bores of different sections, but the problem then lays in the formulation of the non-linearity which, whatever the model, can only be expressed physically in terms of acoustic pressure and flow at the mouthpiece of the instrument.

In this paper, a specific and simple nonlinear synthesis model avoiding the $[p^+, p^-]$ decomposition is used [Gui]. It allows the coupling of the bore with the non-linear characteristics and the reed displacement as it is physically modelled and provides an explicit solution of the coupled nonlinear system of equations in the discrete-time case, avoiding the use of numerical schemes such as the **K**-Method [Bor].

In section 2, we briefly present the classical physical model for the case of a cylindrical bore [Ker2] on which the synthesis model is based, together with its real-time implementation and control. In section 3, this artificial instrument is used in a musical context, in order to study the link between the sound and the performance parameters. Last part is devoted to the conclusions and perspectives of this work.

2 Simplified Physical Model of a Clarinet

2.1 Input Impedance of the Bore (Frequency Domain)

The first linear part of the physical model corresponds to the resonator of the instrument, made of a cylindrical bore. For this geometry, assuming that the radiation losses can be neglected, the impedance relation between acoustic pressure P_e and flow U_e is given by:

$$Z_e(\omega) = \frac{P_e(\omega)}{U_e(\omega)} = i\tan(k(\omega)L) \tag{1}$$

in which U_e is normalized with respect to the characteristic impedance of the resonator: $Z_c = \frac{\rho c}{\pi R^2}$. In the clarinet case, the radius of the bore is large in front of the boundary layers and we define: $\alpha = \frac{2}{Rc^{3/2}}\left(\sqrt{l_v} + \left(\frac{c_p}{c_v} - 1\right)\sqrt{l_t}\right)$. The wavenumber $k(\omega)$ is then expressed by: $k(\omega) = \frac{\omega}{c} - \frac{i^{3/2}}{2}\alpha c\omega^{1/2}$. R is the radius of the bore: $R = 7.10^{-3}$ in the clarinet case. Typical values of the physical constants, in mKs units, are: $c = 340$, $l_v = 4.10^{-8}$, $l_t = 5.6.10^{-8}$, $\frac{C_p}{C_v} = 1.4$.

The length L of the bore is a playing parameter that will determine the pitch. We point out that the internal losses in the bore (dispersion and dissipation) are functions of L and R.

Figure 1 shows the input impedance of the resonator with respect to frequency and its impulse response computed through an inverse Fourier transform. Bore length: $L = 0.5m$, radius $R = 7mm$.

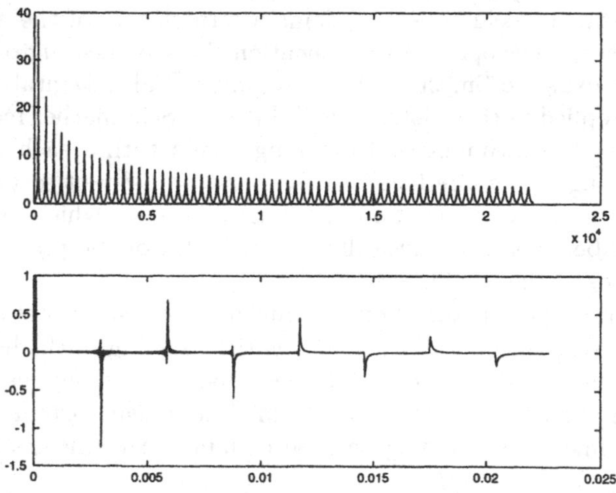

Fig. 1. Input impedance of the resonator (f in Hertz) (top), Impulse response of the resonator (t in seconds) (bottom).

2.2 Simple Reed Model

We use a classical single mode reed model. It describes the displacement $x(t)$ of the reed with respect to its equilibrum point when it is submitted to an acoustic pressure $p_e(t)$:

$$\frac{1}{\omega_r^2}\frac{d^2x(t)}{dt^2}+\frac{q_r}{\omega_r}\frac{dx(t)}{dt}+x(t)=p_e(t) \tag{2}$$

where $\omega_r = 2\pi f_r$ and q_r are respectively the circular frequency and the quality factor of the reed. Typical values for these parameters are: $f_r = 2500Hz$ and $q_r = 0.2$. Writing this equation in the Fourier domain yields the following transfer function of the reed:

$$\frac{X(\omega)}{P_e(\omega)}=\frac{\omega_r^2}{\omega_r^2-\omega^2+i\omega q_r\omega_r} \tag{3}$$

Figures 2 shows the transfer function and the impulse reponse of this reed model ($f_r = 2500\,\text{Hz}$, $q_r = 0.2$, $f_e = 44100\,\text{Hz}$).

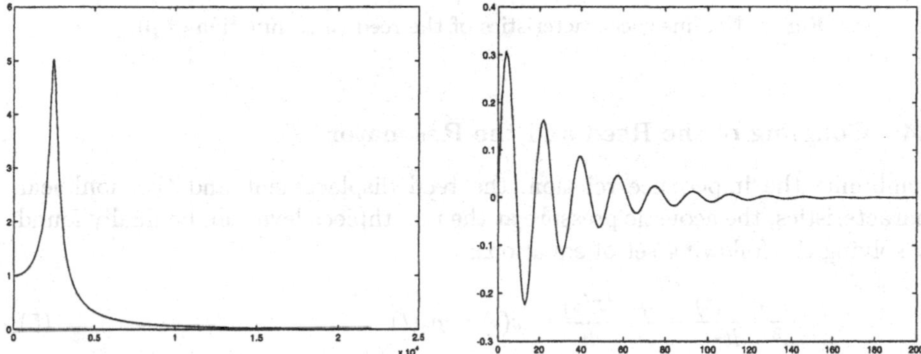

Fig. 2. Left: Transfer function of the reed model (in Hertz), Right: Impulse response of the reed model(in samples).

2.3 Non-linear Characteristics

The acoustic pressure $p_e(t)$, the (dimensionless) acoustic flow $u_e(t)$ and the reed displacement are linked in a nonlinear way at the input of the resonator:

$$u_e(t)=\frac{\zeta}{2}(1-sign(\gamma-x(t)-1))sign(\gamma-p_e(t))(1-\gamma+x(t))\sqrt{|\gamma-p_e(t)|} \tag{4}$$

The parameter ζ characterizes the whole embouchure and takes into account the lip position and the section ratio between the mouthpiece opening and the resonator. It is proportional to the square root of the reed position at equilibrium. Common values lay between 0.2 and 0.6. The parameter γ is the ratio between the pressure inside the mouth of the player and the pressure for which the reed closes the embouchure in the static case. For a lossless bore, it evolves from $\frac{1}{3}$ which is the oscillation step, to $\frac{1}{2}$ which corresponds to the position the reed starts beating. The parameters ζ and γ are the two important continuous performance parameters since they respectively represent the way the player holds the reed and its blowing pressure inside the instrument.

Figure 3 represents the non linear characteristics of the reed for the limit case $\omega_r = \infty$ ($\zeta = 0.3, \gamma = 0.45$). In this case, the displacement $x(t)$ of the reed reduces to the acoustic pressure $p_e(t)$ itself: the reed activity is a single spring.

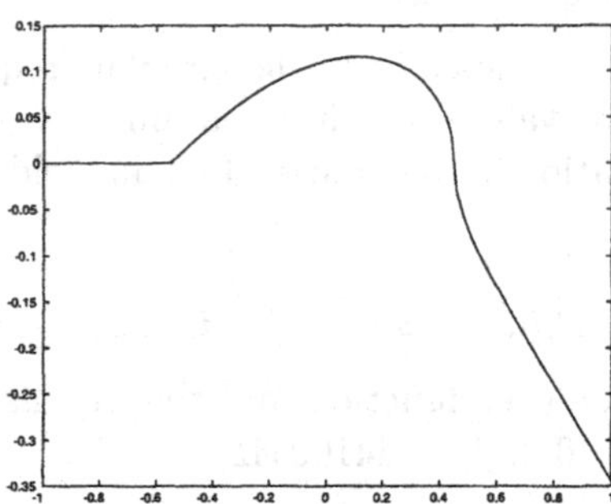

Fig. 3. Nonlinear characteristics of the reed (u as function of p).

2.4 Coupling of the Reed and the Resonator

Combining the impedance relation, the reed displacement and the nonlinear characteristics, the acoustic pressure at the mouthpiece level can be finally found by solving the following set of equations:

$$\frac{1}{\omega_r^2}\frac{d^2x(t)}{dt^2} + \frac{q_r}{\omega_r}\frac{dx(t)}{dt} + x(t) = p_e(t) \tag{5}$$

$$P_e(\omega) = Z_e(\omega)U_e(\omega) = i\tan\left(\frac{\omega L}{c} - \frac{i^{3/2}}{2}\alpha c\omega^{1/2}L\right)U_e(\omega) \tag{6}$$

$$u_e(t) = \mathcal{F}(p_e(t), x(t)) \tag{7}$$

The flow diagram corresponding to this system of three coupled equations is shown in figure 4, in which the reed and the non-linearity are introduced as a nonlinear loop linking the output p_e to the input u_e of the resonator. From the pressure and the flow inside the resonator at the embouchure level, the external pressure is calculated by the relation: $p_{ext}(t) = \frac{d}{dt}(p_e(t) + u_e(t))$.

The model is piloted by the length L of the bore and the parameters ζ and γ of the nonlinearity. It does not require any input signal, since γ is directly proportional to the static pressure inside the mouth of the player. The nonlinearity itself, and its variations imposed by the player plays the role of the source. The digital transcription of equations (5,6,7) and the computation scheme to solve this coupled system is achieved according to the method described in [Gui].

Figure 5 shows the internal acoustic pressure at the embouchure level for a bore of length $L = 0.5$ and radius $R = 7.10^{-3}$. The values of the parameters are: $\gamma = 0.4$, $\zeta = 0.4$, $f_r = 2205$, $q_r = 0.3$. Three phases in the signal are visible: The

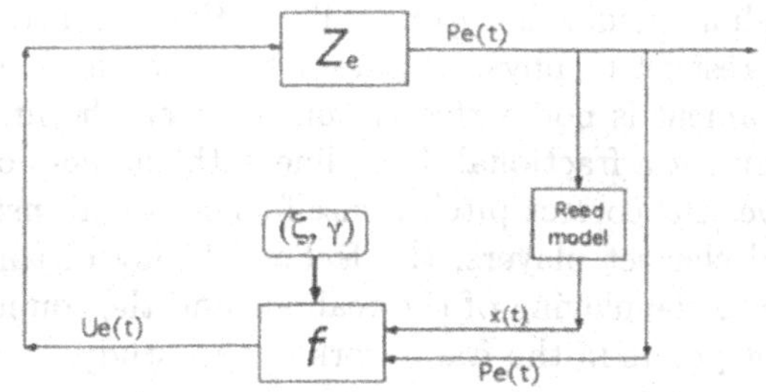

Fig. 4. Nonlinear Synthesis Model.

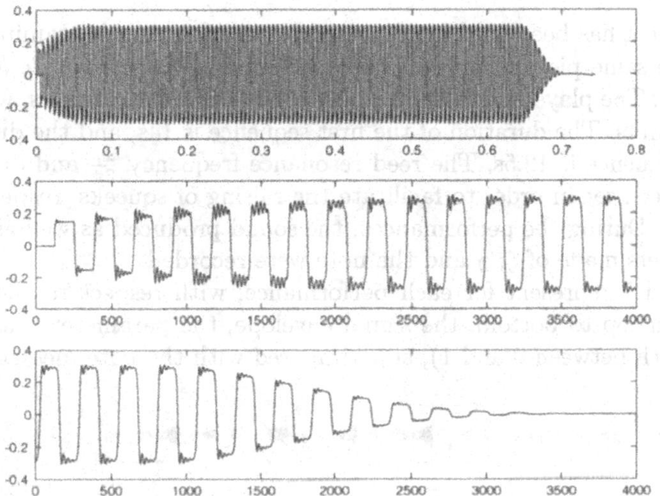

Fig. 5. Top: internal acoustic pressure (in seconds), Medium and Bottom: blow-up of the attack and decay transients (in samples).

attack transient corresponding to an abrupt change of γ and ζ, the steady state oscillations during which γ and ζ are constant, the decay during which γ and ζ decrease slowly linearly towards zero.

2.5 Real-Time Synthesis Implementation and Control

The nonlinear synthesis model has been implemented in real time in the *C* language as an external *Max-MSP* object, piloted from MIDI commands given by a *Yamaha WX5* controller. This controller measures the lip pressure, that controls the parameter ζ, and the blowing pressure that controls the parameter γ. These informations are received in MIDI format (between 0 and 127) in the *Max* patch and are normalized in order to correspond to the range of the physical parameters (between 0 and 1). The tuning of the model is performed by the use of the MIDI Pitch information coming from the fingering which controls the length L of the bore by the relation $L = \frac{c}{4f_p}$, where f_p is the correspondant playing frequency. The delays of the waveguides included in the nonlinear synthesis model are implemented through a circular delay line. Like in the real instrument, since the pitch changes with respect to physical parameters such as γ, ζ, ω_r and q_r and since the real instrument is not perfectly tuned for all the fingerings, it is not mandatory to implement a fractional delay line with the help of an all-pass filter. The player achieves the correct pitch himself as on a real instrument.

According to several clarinet players, the feeling of play of this virtual instrument is comparable to the playing of the real one and the sound is realistic, which are two important points in the framework of this study.

3 Characterizing a Performance Using a Synthesis Model

The virtual clarinet has been used by the same player after a little training, who played twice the same piece of music (The "cat" theme of *Pierre et le loup* by Serge Prokofiev). The player was asked to play of two different manners, without anymore constraints. The duration of the first sequence is 14s, and the duration of the second sequence is 12.5s. The reed resonance frequency $\frac{\omega_r}{2\pi}$ and damping factor q_r have been set in order to facilitate the raising of squeeks, respectively $1850Hz$ and 0.2. During the performances, the sound produced as well as three playing parameters made of ζ, γ and the note were recorded.

Figures 6 and 7 represent for each performance, with respect to the time, respectively from top to bottom, the signal envelope, the parameter γ and the parameter ζ (both between 0 and 1), superimposed with the note messages.

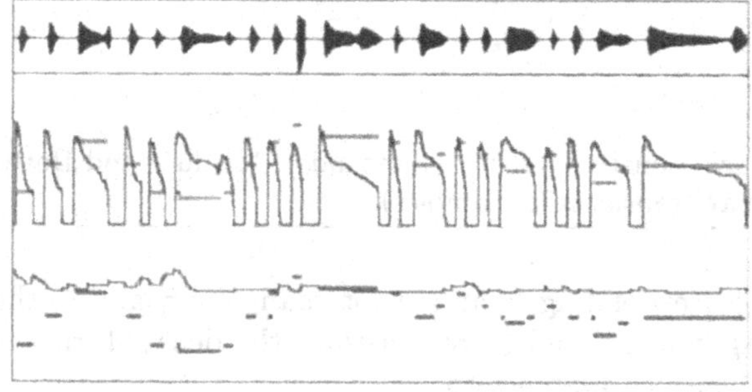

Fig. 6. Sound signal and control parameters of the 1st sequence.

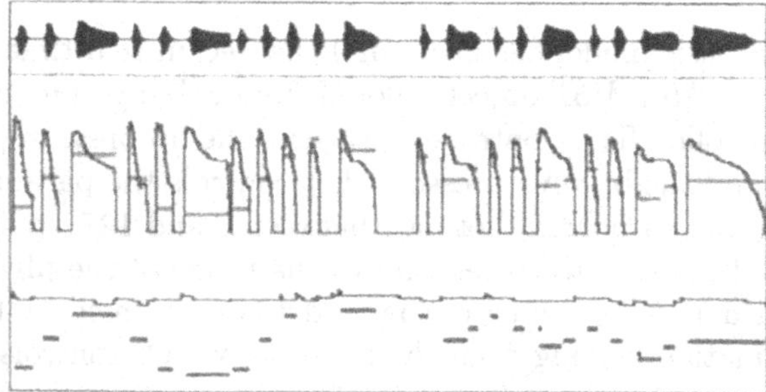

Fig. 7. Sound signal and control parameters of the 2nd sequence.

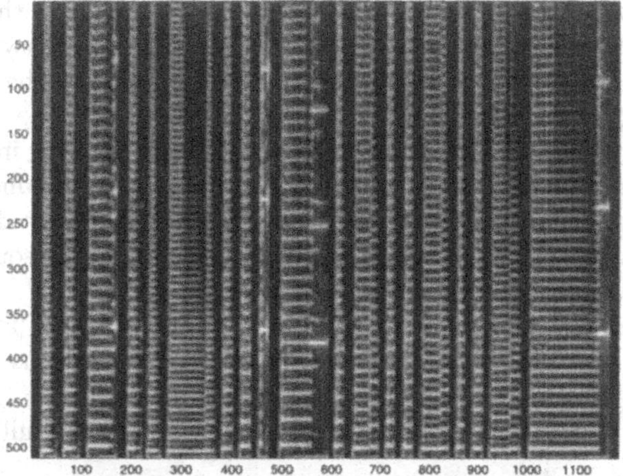

Fig. 8. Spectrogram of the 1st sequence.

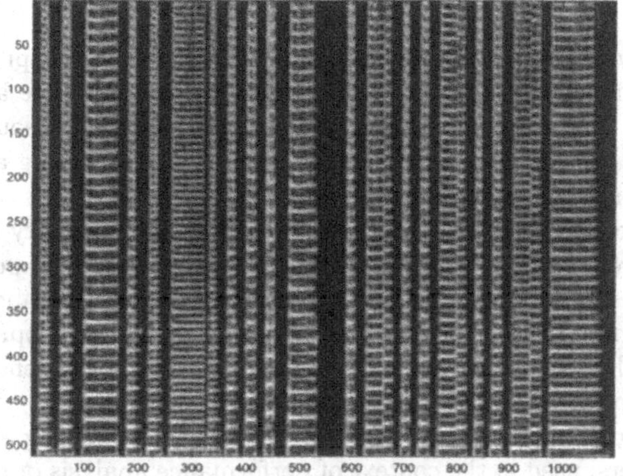

Fig. 9. Spectrogram of the 2nd sequence.

Figures 8 and 9 show the spectrograms (between 0 and 5500Hz) of the two sound sequences (horizontal and vertical scales are meaningless).

For the two performances and for all notes, one can notice that the attack of the sound occurs after the player started blowing. In an orchestral context, this delay has to be compensated. Morover, the blowing pressure is higher during the attack and decreases when the steady state of the self-oscillations is established.

As it is clearly visible on figure 8, the first sequence shows four different squeeks. In this situation, the reed vibrates at a fundamental frequency located around the impedance peaks the frequencies of wich are close to the reed resonance frequency, producing a treble sound with harmonic content. From figure 9, we can deduce that the second sequence is squeek-free. One can notice that the first and last squeeks appear just after instabilities in the way the player holds the reed, together with abrupt changes in the blowing pressure.

Though the variations of the blowing pressure are similar in the two performances, the variations of the sound level and the brightness between the notes played are smaller in the second performance. This seems to be due to more stable values of ζ in this case.

Though the player has been asked to play in a different way in each performance, one can notice that legato and detached notes are the same. In the first sequence, the establishment of the fingering of legato notes can take place after the player starts blowing, but before the previous note starts sounding again, while in the second sequence, it occurs at the same time.

In both sequences, there is an introduction of quick "ghosts" fingerings that do not produce sustained notes, but provide specific transitions between sustained notes.

In the whole, one can say that the player has been more carefull in his second performance.

4 Conclusion

In this paper, we have presented preliminary results of an approach consisting in using artificial instruments in the context of performance and perfomer characterization. The instrument associates a controller and a sound synthesis algorithm based on physical models, in order to obtain a playing and sounding behavior of the instrument close to the real one.

Though the controller we used is simple (it lacks key velocity for example) and far from a real clarinet in terms of touch, aspect, weight, and though the synthesis model is also simple (it lacks radiation of toneholes and vocal tract volume, for example), we showed on an example that this approach can be considered as a light alternative to the instrumentation with sensors of a real instrument and player in the context of performance studies.

Moreover, signal processing techniques aiming at estimating synthesis and control parameters, used in the context of performance analysis or sound modeling for data reduction, can benefit the use of the simple digital synthesis model we used. Indeed, in the clarinet case, the resonator and the reed are each entirely defined with a set of three parameters linked analytically respectively with the radius and length of the bore, and the resonance frequency and quality factor. Such a small set of synthesis parameters, together with two continuous control parameters, might be estimated efficiently using analytic or optimization techniques.

On an other side, collecting simultaneously the sound produced and the control parameters can provide useful datas for the training step of neural networks or other "black box" techniques that can also be used to estimate the playing parameters from sound sequences of real musical instruments.

Further studies in the framework of this article will probably require the preliminary calibration of the artificial instrument, which can by done by the use of an artificial mouth, in order to adjust its behavior to the one of the real instrument the player is used to.

References

[Bol] S. Bolzinger, J.C. Risset, "A preliminary study on the influence of room acoustics on piano performance", in proc. 2ème Congrès Francais d'Acoustique, Arcachon: pp. 93-36 (Editions du Journal de Physique),1992.

[Bor] G. Borin, G. De Poli, D. Rochesso "Elimination of Delay-free Loops in Discrete-time Models of Nonlinear Acoustic Systems", IEEE Trans. Speech Audio Process., 8(5), pp 597-606, Sep.2000

[Duc] Ducasse, E., "Modélisation et simulation dans le domaine temporel d'instruments à vent à anche simple en situation de jeu : méthodes et modèles". Thèse de l'Université du Maine, 2001.

[Gaz] Gazengel, B. , Gilbert, J. et Amir, N. "Time domain simulation of single reed wind instruments. From the measured input impedance to the synthesis signal. Where are the traps?" acta Acustica 3, 445-472, 1995.

[Gui] Ph. Guillemain, J. Kergomard, Th. Voinier, "Procédé de simulation et de synthèse numérique d'un phénomène oscillant", french patent request n0213682, Oct 2002.

[Ker1] J. Kergomard, S. Ollivier, J. Gilbert, "Calculation of the Spectrum of Self-Sustained Oscillators Using a Variable Truncation Method: Application to Cylindrical Reed Instruments", Acta Acustica, 86: 685-703, 2000.

[Ker2] Kergomard J. (1995). "Elementary considerations on reed-instruments oscillations", in Mechanics of Musical Instruments, Lectures notes CISM, Springer.

[Sch] Schumacher, R.T., "Ab initio calculations of the oscillation of a clarinet", Acustica, 48 (1981), pp 71-85.

[Smi] J. O. Smith, "Efficient simulation of the reed-bore and bow-string mechanisms", in Proceedings of the 1986 International Computer Music Conference, The Hague. 1986, pp. 275-280, Computer Music Association.

[Väl] Välimäki,V., and M. Karjalainen,1994. "Digital Waveguide Modelling of Wind Instrument Bores Constructed of Truncated Cones". Proc. ICMC 94, Arhus. Computer Music Association.

[Wal] M. van Walstijn, M. Campbell. "Discrete-time modelling of woodwind intrument bores using wave variables", J. Acoust. Soc. Am, 113 (1), January 2003, pp 575-585.

Conserving an Ancient Art of Music: Making SID Tunes Editable

Gerald Friedland, Kristian Jantz, and Lars Knipping

Freie Universität Berlin, Institut für Informatik, Takustr. 9, 14195 Berlin, Germany
{fland,jantz,knipping}@inf.fu-berlin.de

Abstract. This paper describes our approach for editing PSID files. PSID is a sound format that allows the original Commodore 64 synthesizer music to be played on modern computers and thus conserves a music subculture of the eighties. So far, editing PSID files required working directly with Commodore 64 machine language. The paper gives a small overview of sound synthesis with the Commodore 64, argues why this topic is still interesting for both musicians and computer scientists, and describes our editing approach.

1 Motivation

Many users still remember the Commodore 64 (C64) which took over the market in 1982. It was the best-selling single computer model ever [1]. Jointly responsible for the success were its, for that time revolutionary, music synthesizing capabilities. For creating sounds, the C64 was equipped with a coprocessing synthesizer chip called Sound Interface Device (SID) [2]. Comparable audio quality on IBM compatible PCs has not been offered until years later. Before 1987, when the AdLib soundcard came out, PCs only produced beep tones. As a result, the C64 was used as a standard tool to create sounds for years. Thousands of compositions produced then are still worth listening to even though their sound is quite artificial. Many of todays respectable computer musicians made experiences with the C64[1], like Rob Hubbard[2] who worked from 1988 to 2001 as Audio Technical Director at Electronic Arts. The celebrated Chris Hülsbeck[3], who wrote many titles for C64 as well as Amiga and Atari games, is now composing music for Nintendo Game Cube. SID is a subculture that is still very existing. The High Voltage SID Collection[4], for example, is a large collection of music files. SID music databases are still well maintained, not only because of the unique sound, but also because it reminds people of a culture of their youth [3].

From a computer scientist's point of view, SID files are also interesting because they consist of program code for the C64 that has been extracted from original software. Players emulate the SID and parts of the C64. In contrast to

[1] http://www.c64gg.com/musicians.html
[2] http://www.freenetpages.co.uk/hp/tcworh/
[3] http://www.huelsbeck.com
[4] http://www.hvsc.c64.org

U.K. Wiil (Ed.): CMMR 2003, LNCS 2771, pp. 74–81, 2004.

Table 1. Comparing SID and MIDI

Song	MIDI	SID
Smalltown Boy	51,977	7,601
Das Boot	47,957	4,222
Popcorn	25,486	6,758
Giana Sisters Theme	21,568 (1 song)	23,517 (8 songs)

regular multimedia formats, SID files are computational complete. Converting the format to a modern sampling format or MIDI results in information loss, e.g. endless loops or sounds that depend on random numbers cannot be converted. Editing SID files therefore means automatized editing of a computer program. Modern techniques like flow analysis need to be applied here. The simple architecture of the SID Chip and the C64 makes this kind of music especially suitable to teach the basics of music synthesis and computer programming. The C64 did not have much memory, it came along with only 38911 free bytes of program memory forcing everybody to write economic code[5].

We believe the SID is an important part of computer music history. Our goal is to save this knowledge academically by writing about it and make it accessible for teaching. Giving comfortable editing facilities allows people to experiment with SID sounds.

2 Comparing SID Files with MIDI

Because of the memory restrictions of the Commodore64, SID files are very small. Table 1 gives a comparison of SID and MIDI file sizes in bytes.

Of course the tunes do not sound the same. MIDI files sound less artificial as they are played back with more sophisticated synthesizers. MIDI players also use wave tables to make melodies sound even more natural. A wave table can contain high quality recordings of a real instrument at various pitches. SID was not designed to play sampled sound and wave tables were not used. A composer had to skillfully combine the available waveforms to create the illusion of real music instruments. However, SID creates interesting sound effects and melodies with only a fraction of the size of a MIDI file. Note, that most of the code in SID files shown in the Table has been extracted manually from computer games and may contain graphic control as well as game logic. When playing SID files, this part of the code is redundant and could be eliminated. So the actual code needed to play the sound is even smaller. In both size and quality, SID files are comparable to Yamaha's Synthetic Music Mobile Application Format (SMAF)[6] files used for mobile devices, but SID files are computational complete and can also generate sounds based on calculations. This makes the SID format an interesting choice for storing sound effects. Theoretically it is also possible to create a SID tune that is different every time it is played.

[5] Although experienced programmers knew how to get a few bytes more out of it e.g. by unconventional use of the video memory.

[6] http://www.smaf-yamaha.com

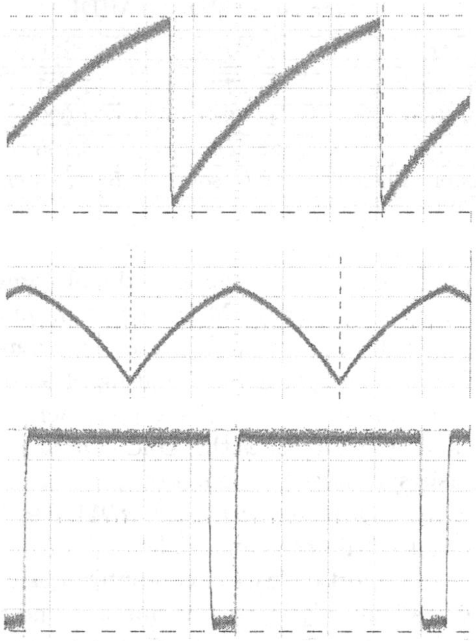

Fig. 1. The SID supported the three shown waveforms and noise

3 Technical View on the SID

The SID chip (MOS 6581) is a synthesizer. It creates tones by overlapping the waveforms sawtooth, triangle, square and noise (Figure 1). The SID is programmed by 28 registers using memory mapped I/O [2]. Music is generated by a regular C64 program that sets values to the registers. A sound is defined by four parameters: pitch, volume, timbre and the volume dynamics. The latter two parameters allow us to differentiate various instruments. The SID chip has three voices that can be manipulated independently. The pitch is controlled by the wave frequency, the volume can be set directly for each voice and the timbre results from the selected waveform. In addition one can create absorbing effects with the help of programmable filters. SID filters have low pass, band pass and high pass filter sections to manipulate frequencies from 30 Hz to 12 kHz. The volume dynamics is controlled by an envelope generator. The generator describes the speed at which the sound rises to his maximal volume and how fast it dies away afterwards. One can also set the volume at which a tone will be held. A musician composing a song was forced to define every single sound on its own to produce the desired song [4]. The C64 User Manual gives tables for looking up the values for waveform and volume dynamics to simulate different instruments as well as frequency tables for different notes. Figure 2 shows an excerpt from a table of frequency values. A composer's task was to put everything together into a program. See Figure 3 for an example. For efficiency reasons, this was usually done in machine code instead of C64 Basic language.

Nr.	Note-Oktave	Frequenz(Hz)	Parameter	Hi-Byte	Lo-Byte
0	C-0	16.4	278	1	22
1	C#-0	17.3	295	1	39
2	D-0	18.4	313	1	57
3	D#-0	19.4	331	1	75
4	E-0	20.6	351	1	95
5	F-0	21.8	372	1	116
6	F#-0	23.1	394	1	138
7	G-0	24.5	417	1	161
8	G#-0	26.0	442	1	186
9	A-0	27.5	468	1	212
10	A#-0	29.1	496	1	240
11	H-0	30.9	526	2	14
12	C-1	32.7	557	2	45
13	C#-1	34.6	590	2	78
14	D-1	36.7	625	2	113
15	D#-1	38.9	662	2	150
16	E-1	41.2	702	2	190
17	F-1	43.7	743	2	231
18	F#-1	46.2	786	3	20
19	G-1	49.0	834	3	66
20	G#-1	51.9	884	3	116
21	A-1	55.0	937	3	169
22	A#-1	58.3	992	3	224
23	H-1	61.7	1051	4	27
24	C-2	65.4	1114	4	90

Fig. 2. Frequency values for certain notes from [4]

```
10 REM MICHAEL ROW THE BOAT ASHORE
20 SI=54272:FL=SI+1:TL=SI+2:TH=SI+3:W=SI+4:A=SI+5:H=SI+6:L=SI+24
30 POKE L,15:POKE TH,13:POKE TL,15:POKE A,3*16+16:POKE H,9
40 READ X:READ Y:READ D
50 IF X=-1 THEN END
60 POKE FH,X:POKE FL,Y
70 POKE W,65
80 FOR T=1 TO D:NEXT
90 POKE W,0
100 GOTO 40
110 DATA 17,103,250,21,237,250,26,20,400,21,237,100,26,20,250,29,69,250
120 DATA 26,20,250,0,0,250,21,237,250,26,20,250,29,69,1000,26,20,250,0,0,0,250
130 DATA -1,-1,0
```

Fig. 3. C64 Basic program for the song "Michael row the boat ashore" from [4]

4 Playing SID Files on Modern Computers

SID tunes were originally C64 programs executed by the computer's 6510 CPU. To make them playable on a modern PC, one needs to extract the relevant parts of the code from the program and transfer it to a PC. This process is called ripping and requires knowledge of C64 assembly language. It is hard work for a "ripper" to decide what is player code and what belongs to the game. He or she only sees assembler code. Because of these required skills there is only a handful of people that convert C64 music into SID tunes[7].

The most popular SID format on the PC is the so called PSID format. It consists of a header containing copyright information, the name of the song, the author, and technical parameters that control different properties like the speed of the tune. The header also defines three memory addresses: the load, init and play address. The load address tells the player where the SID tune is located in the memory of the emulated C64. The init address contains a reference to code executed before playing. The play address is frequently called by the SID file

[7] http://www.geocities.com/SiliconValley/Lakes/5147/sidplay/doc_ripping.html

player to simulate a periodically called interrupt handler which, on the original machine, was used to play the music in the background. If the SID file relies on the init method to play back the music, the call to the play method is omitted. There is a variety of standalone SID players available for example SIDPlay2[8] and SIDPlayer[9]. There are also plug-ins for popular players such as WinAmp[10].

5 The Supporting Framework: Exymen

Exymen [5] is an open source editing tool for arbitrary media formats. It offers an application programming interface (API) allowing to extend the editor for any multimedia format and filtering effect. The extension mechanism is based on an implementation of the OSGi standard [6]. This mechanism, which originally comes from the field of ubiquitous computing, manages loading, updating, and deleting software components from the Internet while the application is running. The concept also permits plug-ins to extend other plug-ins.

Our SID editor is a software component for Exymen, a plug-in that extends Exymen to handle SID files. The SID editing plug-in allows to import the PSID file format into Exymen. Users can edit the files by deleting or creating waveforms, doing cut and paste operations or merging other PSID files. The main advantage of using the Exymen API is that we did not have to implement a new graphical user interface for the PSID editor but can integrate it conveniently into an existing one. Yet, there is a variety of media formats that can be edited with Exymen and hopefully soon to be combined with SID tunes.

6 Editing PSID files

As explained above, PSID files contain two addresses required for playback, the init address and the play address. The playing environment first calls the code located at the init address. It is used to set up the environment by initializing certain memory fields needed later in the program. The init method can also be used to decompress data. Simple decompression algorithms have been used on the C64 already. It is even possible to play back the whole sound in the init method. However, the more popular method to play back SID files is to use the play address. This address is called at some predefined frequency by the SID player. To make editing SID tunes easier, we convert it into a simple event format. The conversion is done emulating the 6510 CPU and listening to the relevant memory addresses that map SID registers. See Table 2 for an overview.

The registers 25 to 28 do not manipulate the sound created by the SID chip. Instead registers 25 and 26 are used to connect input devices and receive their values. Registers 27 and 28 were used by game programmers to get pseudo random values, since they represent the current values of the envelope generator and the oscillator.

[8] http://sidplay2.sourceforge.net/

[9] http://www.uni-mainz.de/~bauec002/SPMain.html

[10] http://www.labyrinth.net.au/~adsoft/sidamp/

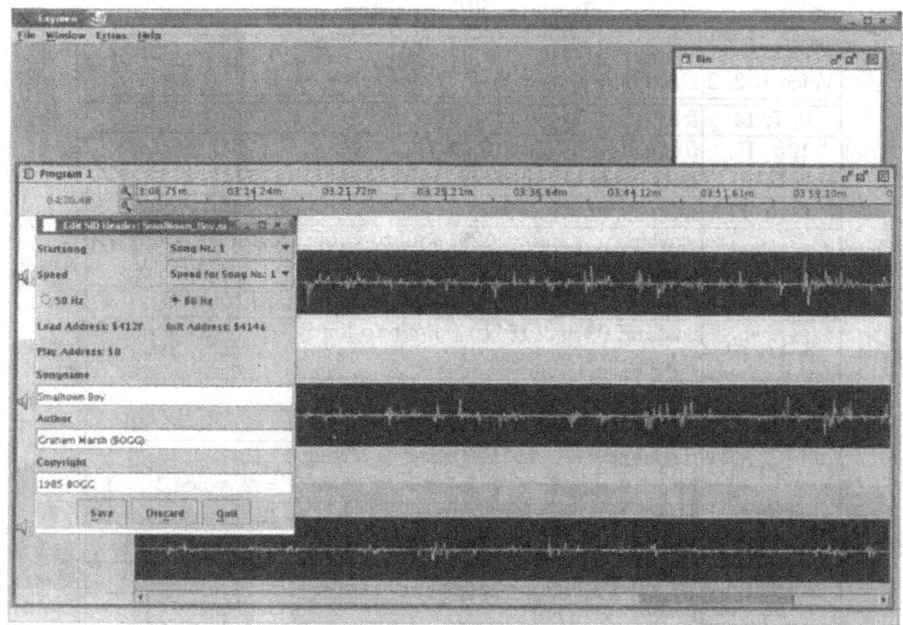

Fig. 4. Screenshot of Exymen editing a SID file

We have decided to use a very simple internal format for a converted SID tune: Every value that has been sent to the SID chip is saved with a time component that measures the distance to the next value and an address. Since the emulator runs faster than the original C64 we use the emulated C64 interrupts to determine the elapsed time between the values. This format allows us to manipulate every single value, since it gives complete control on all data that has been sent to the SID.

The size of a SID file is limited by the small C64 main memory and since a regular SID player depends on this when playing back a SID tune, we are not allowed to exceed this boundary.

Often SID tunes play endlessly because they loop back. The whole SID tune is unrolled by our technique and hence takes more memory to be stored. On the other hand we eliminate redundant code resulting from unclean ripping of demos or games. At the time being we do not detect loops, so we stop converting when 38kB of memory are reached.

The user is now able to edit certain values of the SID file, for example one can copy passages from the SID tune to another thus creating a mix. Another possibility is to modify the time components, altering the speed of the SID tune. One can increase pitch and volume dynamics. Of course, the user is also able to edit all header data (except for load, init and play address) enabling him to modify name and title information.

After editing, the data can be exported to a new SID tune containing a player method written by us.

Table 2. SID Registers

Registers: Voice 1, 2, 3	Description
0, 7, 14	frequency, lo-byte (0..255)
1, 8, 15	frequency, hi-byte (0..255)
2, 9, 16	duty cycle, squarewave only, lo-byte (0..255)
3, 10, 17	duty cycle, squarewave only, hi-byte (0..15)
4, 11, 18	waveform: noise = 129, square = 65, sawtooth = 33, triangle = 17
5, 12, 19	attack: 0*16 .. 15*16, decay: 0 .. 15 (hard to soft)
6, 13, 20	sustain: 0*16 .. 15*16 (silent to loud) release: 0 .. 15 (fast to slow)
21	filter cutoff frequency, lo-byte (0..7)
22	filter cutoff frequency, hi-byte (0..255)
23	resonance: 0 .. 15*16 (none to strong) filter: extern = 8, voice 3 = 4, voice 2 = 2, voice 1 = 1
24	bitmask: high-pass = 64, band-pass = 32, low-pass = 16 volume: 0 .. 15 (silent to loud)
25	paddle X
26	paddle Y
27	oscillator
28	envelope

7 Ongoing Work

The system described here is just the first attempt to make direct and comfortable editing of SID files possible. In order to provide full functionality, SID tunes must also be played back in Exymen. We also want to be able to merge SID music with modern media formats. Another essential goal is to preserve computations in exported tunes.

We believe that we can detect loops with flow analysis to avoid storing data repeatedly. The simplicity of the Commodore 64 is a benefit in this case: Our idea is to keep track of all states the C64 passes while playing back a SID file to notice repetitions. Tracking all states is possible with a modern machine, since C64 programs are small and one state of the 6510 can be stored in 7 bytes[11].

We also plan to use a simple compression in the init method that decompresses the music data before they are played back.

At this moment we are limited to SID files that do without calling the routine at the play address, but produce sound only through a call to the init method. We must adapt our emulator to provide exact timing when calling the play method. This will be done by measuring the CPU cycles between packets.

[11] Of course, using this analysis relies on SID tunes not using self modifying code.

8 Summary

With the system presented here, an interesting piece of computer music history is conserved for teaching purposes. The experiences with the SID, the knowledge of C64 and the basics of music synthesis is of unpayable value. The SID technology introduced many of todays computer music professionals into their field. Music on the C64 is still an active sub culture. Not only are there still many fans of SID music, there is even a company that sells the SID chip as a PCI card for modern PCs[12].

The topic is interesting for teaching, because it offers opportunities to use modern techniques like compression and flow analysis on a simple but real architecture. The size of the tunes compared to their quality is really remarkable. We think the technical and artistical achievements of former times should not be forgotten.

The project started in a multimedia course at Freie Universität Berlin and has been released open source[13].

References

[1] G. Friedland, Commodore 64, in R. Rojas (ed.) Encyclopedia of Computers and Computer History, Fitzroy Deaborn, New York 2001.

[2] PATENT NO.: 4,677,890, ISSUED: July 07, 1987 (19870707), INVENTOR(s): Yannes, Robert J., Media, PA (Pennsylvania), US (United States of America), (APPL. NO.: 6-455,974 FILED: February 27, 1983 (19830227).

[3] Mathias Mertens, Tobias O. Meißner, Wir waren Space Invaders, Eichborn, Frankfurt 2002.

[4] Commodore Business Machines, Commodore 64 User Manual, 1982.

[5] G. Friedland, Towards a Generic Cross Platform Media Editor: An Editing Tool for E-Chalk, Proceedings of the Informatiktage 2002, Gesellschaft für Informatik e.V., Bad Schussenried 2002.

[6] OSGi Service Platform Release 2 (2000), edited by OSGi, IOS Press, Amsterdam 2002.

[12] http://www.hardsid.com

[13] See http://www.exymen.org

"GMU" - An Integrated Microsound Synthesis System

Laurent Pottier

Research engineer - Gmem (Marseille)
Centre National de Création Musicale

Abstract. The GMEM (National Center for Musical Creation) has created an integrated microsound synthesis system called GMU ("GMEM Microsound Universe") inside the Max/MSP environment. GMU works in real time and its center is made of a few microsound synthesis objects (written in C at GMEM). Around these objects, we have elaborated sophisticated sound control strategies, with high-level interactive access inputs. They can be deterministics or stochastics. The whole system offers many connections with tools for Composition Aided by Computers (written in Common Lisp and Java) and tools for sound analysis. Before describing our system and its applications, we will show how the synthesis and transformations of microsound have already been significant in music nowadays.

1 Introduction

Sound synthesis made by computers to create music has undergone very important developments in the late sixties and seventies. Our knowledge about the nature of sound has vastly improved during this period

At the end of the late sixties, Max Mathews had already enounced the essential of the sound synthesis techniques known today [4]. Jean-Claude Risset described the way to synthesize most musical instrument sounds [8].

At the end of the late seventies, Xavier Rodet proved that it was possible to produce singing voice synthesis by using a computer. He found out that the main difficulty is to develop parameter control convincing rules. In that, synthesis parameters must be interconnected [12].

During the late nineties, sampling expanded a lot, reducing the use of sound synthesis in music. Actually, the development of microcomputers and home studios has promoted the use of traditional sounds (with the General Midi standard). This has slowed down research of new sounds.

Recently, a lot of people have become more and more interested in new sounds. Some open environments, like Max/MSP, let the user construct his own synthesis tools. He needs to do programming, which is sometimes tricky, because he has to take into account many constraints related to "real time" processing. Other environments, like Csound, belonging to Max Mathews "Music N" programs, are used to program instruments which can produce sound "out of time". The user can concentrate only on producing sound, without the real time constraints.

U.K. Wiil (Ed.): CMMR 2003, LNCS 2771, pp. 82–88, 2004.

2 Microsound Synthesis

What we decided to do is consider sound as a particle process rather than as a wave process.

This is not a new idea but generally it is not used in the main classical sound synthesis techniques in which harmonic analysis is the cornerstone. Gabor was obviously the first to have defined the notion of "sonic quanta", which is a very short sine wave [3].

Then Xenakis has developed a new composition theory in which he takes grains as the basic constituent elements. He proposed to create complex sounds with random distributions of thousands of grains, that he called a "cloud".

> "All sound, even all continuous sonic variation, is conceived as an assemblage of a large number of elementary grains adequately disposed in time?. In fact within human limits, using all sorts of manipulations with these grain clusters, we can hope to produce not only the sounds of classical instruments and elastics bodies, and those sounds generally preferred in concrete music, but also sonic perturbations with evolutions, unparalleled and unimaginable until now. The basis of the timbre structures and transformations will have nothing in common with what has been known until now" [16]

He proposed different stochastic laws about the control of the parameters of grains. This allows one to get closer to the natural sounds without having to specify the time evolution of each grain.

Since the end of the seventies, Curtis Roads has built different programs to generate artificial sound grains. He tried to create a granular sound system, with some controls of the sound parameters, allowing optimal sound quality and the best flexibility for a musical use. While studying the shapes of the amplitude envelopes of the grains, he has proposed a good compromise for these shapes: he uses Gauss functions for attack and release, and a small steady state in the middle [9].

In the late eighties, Barry Truax was interested in synthesis techniques that would allow him to create sounds similar to environmental sounds. He achieved one of the first and most powerful granular sound synthesis real time system (PODX on a PDP11 computer) [14].

In 1994, I worked with the composer Manuel Rocha. He made many pieces using only granular sound synthesis. For him I created many tools inside "PatchWork", the Ircam algorithmic composition program, for the control of granular sound synthesis made with Csound [6].

Later on, Manuel Rocha built, with the computer programmer Gerhard Eckel, the GIST "Granular Synthesis Tool Kit" system on the Ircam Musical Workstation. This system generalized the synthesis sound technique of the FOF ("Fonctions d'Ondes Formantiques", used for singing voice sound synthesis) for the micro-fragmentation of sampled sounds [1,11].

3 The GMU Project

Considering these different concepts, we have decided at Gmem to create our own microsound synthesis tools and to put them into a more general integrated compositional system for control.

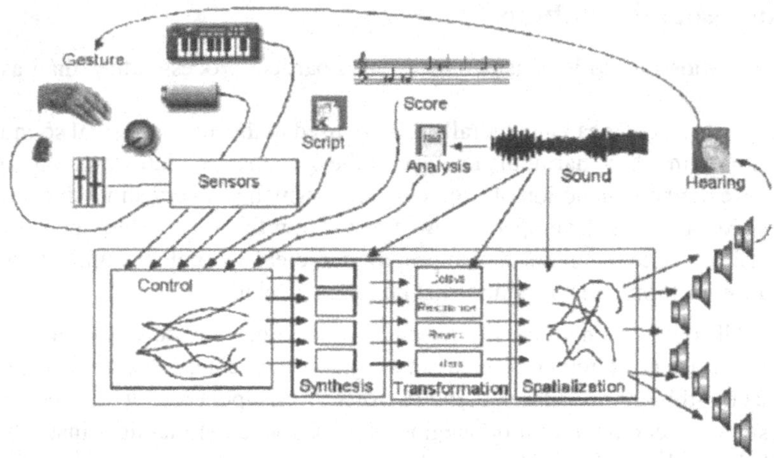

Fig. 1. The GMU microsound synthesis system.

This system has to take into account the different steps and methods involved in the compositional process. At the same time, we propose powerful sound synthesis generator units, a multi-speaker holophonic system and models and tools for the parameter control.

This system can be used for real time interactive applications on Max/MSP platform and for non real time applications on Csound. In this text, we will discuss only the Max/MSP version.

4 Generator Units

The base of our environment is made of several microsound generator units (grain makers) programmed by Loic Kessous (PhD student - Paris VIII University) and myself. These units create polyphonic streams made of grains with which waveforms and amplitude envelopes can be of several types.

The Envelope. Three types of grains envelopes are available:

- linear envelopes made of three straight lines of variable relative durations,
- gaussian envelopes,
- envelopes with a sinusoidal attack, a sinusoidal decay and a steady state.

The linear envelope contains some discontinuities making the sound aggressive and rough, especially in asynchronous streams. It can be used to produce streams where grains overlap and then can generate continuous partials (providing that phases are respected).

The gaussian envelope is very soft but most of the time it doesn't contain enough sound signal. It is useful with high-density clouds.

Lastly, the third envelope, which can be found in the FOF synthesis, is a very good compromise to create many types of sounds. It contains both a steady state and soft attack and decay.

The Waveform. Three situations have been considered:

- the waveform is artificially synthesized,
- the waveform is taken inside a memory stored sound,
- the waveform is taken from a live play sound.

In the first case, we can use a sine waveform but it is possible to use many waveforms stored in tables. It is possible to apply non-linear distortions to a sine wave (frequency modulation or another technique) too. Now, only the sine waveform is available in our real time system.

In the second case, the grains are taken from recorded sounds that can be instrumental, environmental or others sounds. The microsound synthesis can then resynthesize the original sound if the grains are played in the right order and if some criterions are respected. The grains can be played with different speeds and different rates so we can create transpositions or duration variations of the original sound. It allows the user to observe the sound as if through a microscope. Barry Truax has sometimes realized sound stretches with a factor of 100.

Other Parameters. Considering one type of envelope and one given waveform, a grain is characterised by the following parameters: its duration (typically from 10 to 50 ms), the transposition (or the frequency), the attack and decay's lengths, the position in the sampled sound where we take the grain (for sampled waveforms only) and the output direction. Inside a grain, the parameters don't change.

Start Times. The different grains are triggered by an audio signal. One grain is triggered when the audio signal goes from negative values to positive values (zero crossing).

If we use a sine waveform to trigger the grains, we obtain periodic triggering.

Several grain makers can be used at the same time and be triggered in a synchronous manner (with the same audio signal) or asynchronous. A synchronous triggering can produce harmonic sounds. This is the technique used, for example, for a singing voice synthesis. Asynchronous triggering can produce complex textures. The timbre of these sounds depends on the way we change parameter values.

The synthesis needs to produce streams composed of thousands of grains each second. The grain makers that we use have to produce up to one hundred grains simultaneously.

"Grain makers" have a lot of outputs. Panoramic parameter tells us which output will be given to each grain. The grains are sent to resonators and reverberations that contribute to the timbre of the sound.

5 Control of the Parameters

A grain maker can be controlled directly by sending values to it. These values can be produced by stored messages or by a gesture input device.

At a higher level, some functions have been programmed to deliver stochastic values (Brownian movement) between a minimum value and a maximum value.

For sound production to varies in time, the ranges can be modified dynamically. A graphical editor allows the user to draw curves describing the evolution of each parameter in relation with time.

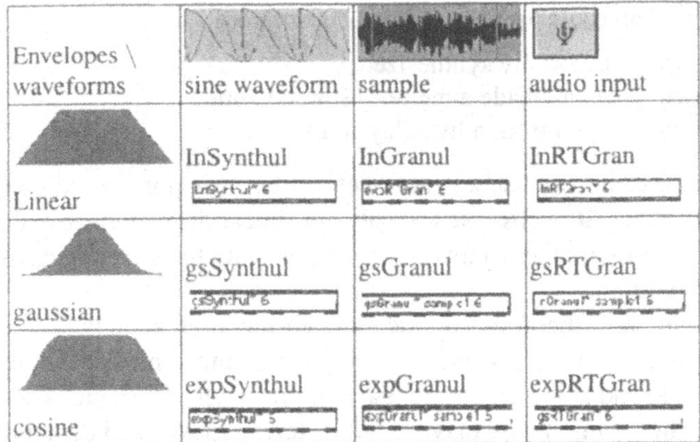

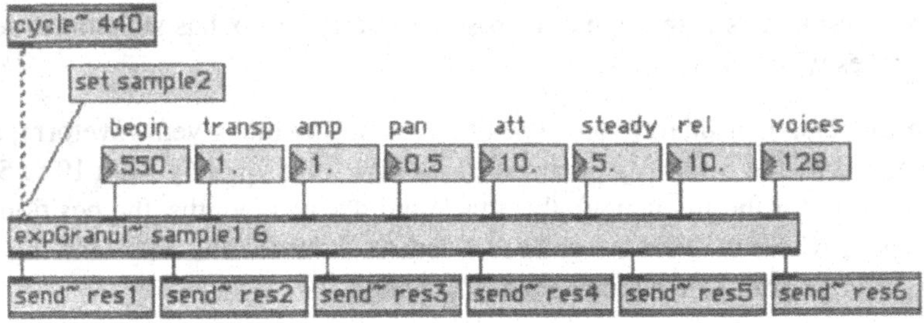

Fig. 2. The expGranul obejct with its synthesis parameters. Separated outputs is triggered with a sine signal.

Finally, connections were made by Jean-François Oliver (ATIAM DEA student) and myself with programs for algorithmic composition (Holo-Edit from Gmem or Open-Music from Ircam), which makes it possible to develope the evolution of the various parameters with graphics, music notation, sound analysis or many algorithms,

Some tools were realized in Max/MSP/Jitter to graphically represent the parameters of grains. They were made with the Open-GL graphical libraries. They help to control results of the synthesis. In future applications, they could be used to build artistic illustrations of sound.

6 Applications

Our main goal is to produce tools for musical applications. We wish to remain connected with musical production and to adapt our tools to artistic demand.

The first musical experiences I made with microsound synthesis were with Manuel Rocha ("Transiciones de fase": 1995) and Jean-Baptiste Barrière ("Le Messager": 1996) with the GIST synthesizer on the Ircam Musical Workstation.

In Marseille, during the festival "Les Musiques-2002", we used our system for the piece by Eric Abecassis "Straps" for the real time granulation of acoustic instruments played live.

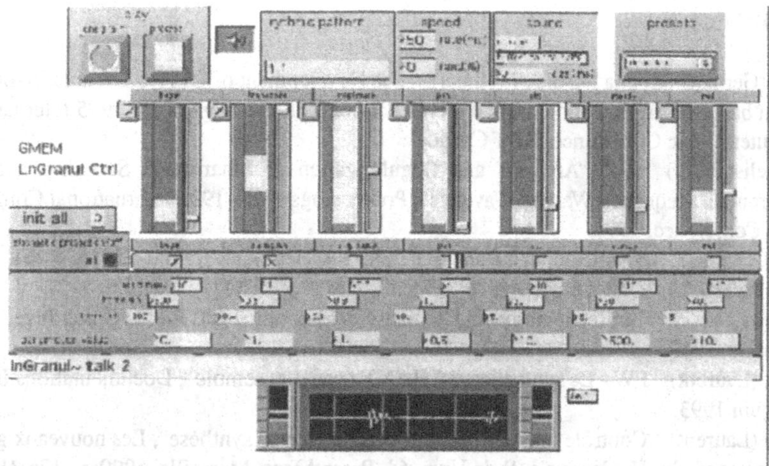

Fig. 3. Max patch with sliders used for control of range inside which the parameters for grains synthesis are calculated.

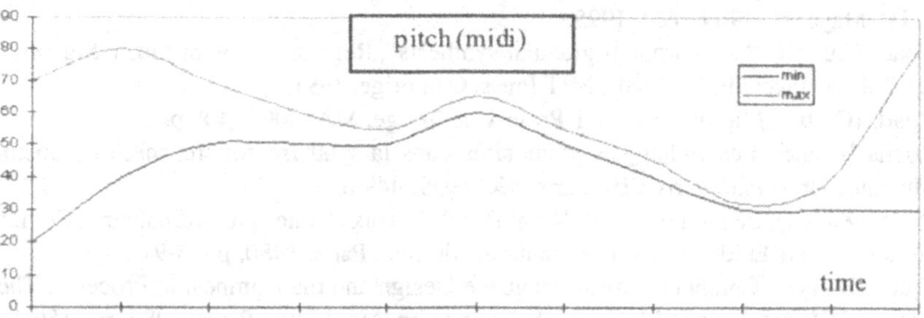

Fig. 4. Graphical description of the variations of pitch of grains.

At Gmem, composers have always been attracted to environmental sounds more than to harmonic sounds. Their main tool is the microphone as a first step and then signal manipulations inside a studio. Granular sound synthesis techniques interest them more and more because they have the advantages of both synthesis and sound processing. Primarly, these tools are built for fulfill their needs.

7 Perspectives

We are certain that synthesis technology is promising. Contrary to the widespread opinion observed in the late nineties, we think that all possibilities available with sound synthesis have not yet been explored. Sound synthesis is a very new field in music, especially with the exploration of micro-time domain, at a scale lower than perception. Microacoustic phenomena has been badly understood as yet. We must explore this domain and develop new tools and new expertises.

References

1. Eckel (Gerhard), Rocha Iturbide (Manuel), "The Development of GIST, a Granular Synthesis Toolkit based on an Extention of the FOF Generator", Proceedings of the 1995 International Computer Music Conference, Banf Canada.
2. Evangelista (G.) 1997, "Analysis and Regularization of Inharmonic Sounds via Pitch-Synchronous Frequency Warped Wavelets", Proceedings of the 1997 International Computer Music Conference, 1997.
3. Gabor (Dennis),"Acoustical Quanta and the Theory of Hearing", Nature, 159, n° 4044, 1947, pp. 591-594
4. Mathews (Max), The Technology of Computer Music, éd. MIT Press, Cambridge, Massachussets, 1969.
5. Pottier (Laurent), "PW - Le synthétiseur CHANT : vue d'ensemble", Documentations Ircam, Paris, juin 1993.
6. Pottier (Laurent), "Contrôle interactif d'une voix chantée de synthèse", Les nouveaux gestes de la musique, dir. H. Genevois, R de Vivo, éd. Parenthèses, Marseille, 1999, p. 175-180.
7. Puckette (Miller), "Combining event and signal processing in the MAX graphical programming environment", Computer Music Journal, vol. 15, n°3, 1991, p. 68-77.
8. Risset (Jean-Claude),"An introductory catalogue of computer synthesized sounds ", The historical CD of Digital sound synthesis, Computer Music Current 13, éd. Wergo, WER 2033-2 (CD), Mayence, 1969, rééd. 1995.
9. Roads (Curtis), "Asynchronous granular synthesis", Representation of Music Signals, dir. G. De Poli, A. Piccialli, C. Roads, MIT Press, Cambrige, 1991.
10. Roads (Curtis), Microsound, MIT Press, Cambridge, MA,:2001, 408 p.
11. Rocha Manuel, Les techniques granulaires dans la synthèse sonore, thèse de doctorat en Musique, Université Paris VIII, Paris, déc. 1999, 449 p.
12. Rodet (Xavier), Bennett (Gerald)," Synthèse de la voix chantee par ordinateur ", Conférences des Journées d'Etude, Festival International du Son, Paris, 1980, p. 73-91.
13. Truax (Barry), "Computer Music Language Design and the Composing Process", The Language of Electroacoustic Music, éd. S. Emmerson, MacMillan Press, 1986, pp. 156-173.
14. Truax (Barry), "Real-time granular synthesis with a digital signal processor", Computer Music Journal, 12(2), 1988, pp. 14-26
15. Truax (Barry), "Time-shifing and transposition of sampled sound with a real-time granulation technique" , Proceedings of the 1990 International Computer Music Conference, Glasgow, 1990, p. 104-107.
16. Xenakis (Iannis), Formalized Music, Bloomington, Indiana University Press, 198-71.
17. Xenakis (Iannis), "More through Stochastic Music" Proceedings of the 1991 International Computer Music Conference, 1991

What Are We Looking for?

Daniel Teruggi

Composer, Research Director
and Director of the GRM (Groupe de Recherches Musicales)
at the Ina, (National Audiovisual Institute)
Bry sur Marne, Paris, France
dteruggie@ina.fr

Discovering information is one of the great challenges of our time. Description tasks are among the most time consuming activities in our society and we leave behind huge swathes of documents, which we will never describe or address due to lack of tools and time. At the same time humankind has produced and will continue producing unimaginable quantities of organised information through different media (photographs, books, music, television, radio, internet) based on the hypothesis that there is a potential viewer, listener, reader or navigator that will make sense of it through their perception.

A close relation has developed between scientists and psychologists in order to understand and model our mind's capacity to interpret information and to make sense of it under the best and most effective conditions. But the objective seems always beyond our grasp; its is not computer power we are lacking, it is an understanding of the way we construct meaning in order to build intelligent systems capable of improving the automation of many tiresome and repetitive activities.

At the same time However, it is a highway for research. We are working on understanding how we function. From there on, we build systems that simulate behaviours, then test them in order to measure the gap between reality and virtuality to then build on new models and methods. Little by little, our understanding advances and we conceive better and faster algorithms to cope with this understanding. However, there is always a wide gap between signal and concepts -low-level and high-level as they are called. Signal processing is at the low-level end, high-level concepts in our brains are at the other. There are few shortcuts going from one end to the other, capable of making our systems accomplish those high-level tasks that seem so easy for our mind, through intelligent choices and general functioning models. Statistical and comparative methods are the best tools we have been able to develop up to now, but they seem primitive and awkward when it comes to doing tasks that seem easy for our mind, such as following a melody in a complex harmonic environment.

1 Using the Human Being as a Model

We have always assumed that an intelligent system would have to use the same methods used by our brain in order to understand information. This may or may

U.K. Wiil (Ed.): CMMR 2003, LNCS 2771, pp. 89–97, 2004.

not be true, in any case, it seems the most reasonable precept on which to build models of perception and apply them to recognition tasks. So, for quite a long time we have being working on signal processing in order to build algorithms capable of identifying, recognising, measuring and then extracting conclusions from a signal. On the other hand, we have been trying to understand how the brain functions and how we make precise and fast perceptual choices in order to extract meaning in an economic and effective way.

The perception process that we actually perform in our everyday life looks something like this:

1. We isolate information from our surrounding environment and concentrate our attention on the particular channel or media conveying it, thus focussing on a particular perception organ. We identify the fact that there is organised information underlying the channel and we concentrate our activity on the particular aspects of that organised information, thus separating different kinds of information or information from noise.
2. We apply the most effective method of extracting meaning from the perceived information using a global approach to establish both a perceptual framework and to make a fine analysis of the details. We organise the perceived information in structures in order to simplify the comprehension and deduce possible meanings for it.

The two main concepts in this model are that we are capable of isolating information from a very disorganised context in order to interpret it, and that then we use the best-adapted method to guarantee comprehension.

These tasks may be performed in very simple situations with only one incoming flow or in extremely complex configurations, in which different and contradictory information coexists at the same time. This search for information can be free or oriented. Free when looking for any kind of information or organisation within a media, just being alert; and oriented when we focus the research in order to find a particular kind of information.

The human model is a good one since it is efficient, fast and known to us. Making a system function as our minds do has seemed a good approach in the past but now we know that the simplest tasks are the most difficult to achieve. It seems reasonable to continue investigating the brain model option as well as advancing in other paths that may lead to new solutions. This is a general artificial intelligence problem, and we should keep this in mind when addressing the musical domain with its particular problems and interpretations.

2 How to Understand Music

If we analyse the state of Art, we can say that these processes have been quite successfully applied to the interpretation of language. The so-called intelligent systems can identify morphemes, associate them to words and sentences and deduce possible meanings by mapping all the possible meanings of a sentence and then proposing the most plausible interpretation, sometimes in relatively noisy situations. To arrive at such a result it has taken the cognitive sciences years to develop models that describe the behaviour of our mind in the process of constructing meaning. The major assumption in this approach is that there exists a general common structure for languages and it can be represented through grammar, which is a distant model of how our brain functions.

If we could have similar systems applied to music, what would we ask them to do? It would indeed depend on which music we are talking about. There have been in the past more or less well defined organisational rules, which were relatively prescriptive in different historical periods. These rules summarised the general practices of the time and up to some point assured an acceptable result if applied correctly. There were also prohibitions, to guarantee that music would not allude to "forbidden" references (the Major second recalling Arab music in the Middle Ages for example).

Music rules, and mainly the rules known as the tonal system, were constantly evolving and adapting to the uses composers made of them. In addition, since rules were only ways for composing and not objectives, composers kept stretching the limits of the rules, thus making the whole system evolve. There was of course no relation between the good use of the rules and the quality of music, but some composers came to be seen as prototypes for rules of a certain period, while others cannot be completely analysed within a rigid set of rules. The main aspect of any rule-system in the past, was that they were a consequence of the uses and in a certain way tried to summarise the main trends underlying the music of a composer and eventually of a period.

In the twentieth century, there was a remarkable change: sets of rules did not preordain music making, and were no more a framework on which the composer would work to organise his ideas and intentions. The progressive expansion and disaggregation of the traditional tonal model did not open the door to disorder or arbitrariness, it opened the way to new rule systems that permitted new sound organisations with specific perception frames. While doing this, the new systems had to fight against the strength of the major rules of the former tonal system and the major tendency for a listener of the time to interpret a "dissonance" as an exception within the well known tonal system.

Little by little ears evolved and acceptance improved. It became a fact that music could exist under different and contradictory rule-frames, and that composers had to organise the sound material before composing since there were no more underlying structures shared by the composer and the listener. Listeners (when eager to approach new music), learned very quickly that it was possible to listen and appreciate music even if the rules were previously unknown and that they could build a perception framework for understanding and enjoying a new musical experience. This also poses a new aesthetic problem as to how beauty can be perceived if we do not know the rules of beauty, but modern art has extended this problem to all the arts and it can be conceived that pleasure is possible with no previous knowledge of the framework.

There was a tremendous break with the arrival of *Musique Concréte* and electronic music. Up to then, even if the rule-systems changed, there was a continuous and coherent aspect in music making, and this was the sound material: musical sound had not made major changes during almost three centuries. Sound combination had progressively become more complex and instrument performance had permitted new sounds, but the basics of music making had not changed during the first half of the twentieth century (even if attempts to create electric instruments existed). Radio and other new media, opened the door to the use of any sound in music, thus bringing new perception problems, mainly understanding what underlies the act of perceiving a succession of sounds as being coherent and homogeneous.

Musique Concréte began with no well designed rule-system (this is the main difference with electronic music and its use of serial techniques in order to attain a perfect control on all sound parameters). Pierre Schaeffer understood quite quickly the limits of a non-ruled system and began to study the possible underlying rules that made music be musical with these new techniques. Experimentation became one of the key aspects in the work done by his Group and he would constrain composers to conceive rule-systems, or to apply theoretical procedures during composition with the intention of obtaining a general grammar that would be efficient for any kind of music. His main achievement was the *Musical Objects Treaty* (Traité des Objets Musicaux) published in 1966, in which he approaches the understanding of music through an analysis of perception.

In a certain way, while trying to understand what made music be music, he wanted to understand how a composer composed music. Many years later, when he had lost interest in the music he had helped develop and gone back to his first musical references of the eighteenth century, he would demand that young Conservatory composers explain the underlying rule-set before listening to any music.

What happens today, when so much music ranging from ancient to contemporary, from ethnic to popular, is available for our listening? Composing is an open adventure, in which composers organise a set of rules or a framework through which they develop their ideas and conceptions. The total world of sounds available for music is so wide that it has to be reduced to a selection; the possible organisation rules are so varied, that reduced sets are used in order to assure coherence and understanding of the functioning.

Because the main problem remains the listener; music is intended for listening and there is a constant concern by composers of any kind that their music should be understood and that it will interact with the listener, through pleasure, movement or intellectual delight. In the beginning there were few sounds and simple rules, today there are millions of possible sounds and rules but only a few of them are normally used in a work, in order to keep complexity and diversity within the limits of perception.

3 What Should an Intelligent System Look for Then?

This is the main question: faced with such a diversity of sounds, rules (grammars) and situations, how can we teach a system to make its way through listening? Are there common trends to all music, so global and wide that we could model them?

This short history of rule-systems aims to show that it is very difficult to identify a common pattern to all music. There could be a temptation to associate musical styles with languages, but if English and Japanese can be considered as completely different; they share common structures and common functions for the words they use. There are no common rules between plainchant and Techno music, except the fact that they both address our senses and induce in us the perception of an organised structure.

Maybe before answering these general questions, we should try to define in what way would an intelligent system be useful. Do we want systems to perform tasks we find tiresome and repetitive, or do we think that with new tools we will be capable of performing new actions and getting new understandings of the musical phenomena? Developing tasks in music implies working on two possible ways of representing it; through a written interpretation with a score and notes (and a MIDI transcription), or through an acoustic description that we address through listening or with the signal processing of a recording.

Several domains can be identified, within which some important research has been carried out and very practical tools created. These simplify repetitive tasks and are capable of extracting information:

- *Music Identification:* one of the major outputs for recognition, how to recognise an author, a theme or vertical organisation through comparison with a reference database. It has important applications in the domain or rights management and for automatic documentation and classification. Sound identification can also be used for security applications.
- *Classification and retreival:* a major problem in sound production and music management due to the huge amounts of material to classify. One of the main challenges involves classifying items through content analysis and retrieving them through signal defined queries. This domain implies the construction of sound an music ontologies to organise the information in coherent description systems. Classification and retrieval have strong applications in music and cinema production and post-production.
- *Structure extraction:* Searching for organisation within a document. Any kind of regularity, repetition is tracked in order to identify structures, themes or patterns that permit a description of the document as a whole. It is mainly applied to scores and is beginning to be applied to signal.
- *Sound spotting:* Looking for specific information within a document or a database, it can be speech, sound or note oriented. This implies not only the search for identity but also the capacity of identifying sounds or music in different contexts as well as variations of the original (pitch, duration, spectral or instrumental).
- *Statistic studies:* Done mainly on scores, in order to study, compare or extract musical patterns through its note description. Difficult to apply to signal processing due to the lack of discreet units. Quite efficient for extracting style patterns of a composer or eventually for interpolating melodies.
- *Musical analysis:* One of the major outputs for future applications, very little analysis assistance tools exists. The main problem in analysis is the

variety of points of view an analyst may apply while studying music. Existing statistic and structure extraction tools may be applied but are not dedicated to analysis. A strong effort has to be made to formalise the analysis tools and to adapt them to different possible uses.
- *Sound tracking:* Following within a musical pattern a particular instrument, melody or sound contour. Applied to signal processing it permits a predefined line to be followed or to identify polyphonic patterns through spectral differences.

The above is a list of some useful actions that systems can perform on sound and music, mainly for practical purposes and to assist us in complicated and boring tasks. We are not working with high level concepts, as for example wanting to understand if a work is beautiful or why it is boring! Today's tools help us acquire a comprehension of the components that will permit us to infer why we find a work beautiful or why it seems so boring. However, we can also have these higher ambitions in the future and ask an intelligent system to explain the structure, the elements, the beauty, the way the composer develops his ideas, his style or to identify his ideas and trends.

4 What Makes Music Be Music

An astonishing conclusion for our perception of music is that we accept different sound organisations, covering a thousand year span, representing cultures of the whole world and with different levels of complexity as all being part of the same phenomena. This is maybe an intellectual conclusion, since I am conscious of history, geography, society and I can easily identify dozens of examples of music that seem radically different one from another. Maybe listening to music is a much easier task than what my perception and intellect deduce and there are finally strong common trends between any music that make its tracking and interpretation easier than what one could imagine.

In such a search for simplicity and common trends, I would point out several aspects that seem evident to any kind of listening and are associated with perception and musical listening practice:

- *Regularity:* it is the first clue that may indicate that some kind of organisation exists. Traditionally, meter created a framework for regularity but it may be completely disguised. Other forms of regularity exist, through sounds, intervals or regular patterns. Probably the strongest hint to identifying if a sound phenomenon is musical.
- *Homogeneity:* whatever the music be, electroacoustic or instrumental, there is a certain coherence in the sounds used and in the way they are used. Astonishingly enough, in electroacoustic music where extreme diversity is very easy to obtain, composers define a relatively limited set of sounds on which to work. Listeners try to understand as fast as possible the set of sounds underlying a particular music.

- *Variation within regularity:* extreme regularity may be easily associated with a machine or a non-human functioning. Detection of irregularities that suggest a human behaviour or control, even in electroacoustic music, is a very strong hint of musicality.
- *Repetition:* probably a generic aspect of all music is a strong amount of redundancy, in a short period level it deals with metrics and rhythm; in longer periods it deals with themes, phrases and structures.

We could probably add more trends that would suggest musicality, for instance the formal evolution through time, spectral regularity, harmonic distribution (several similar sounds happening in the same instant) or complexity (neither too many nor too few events). No trend is strong enough to convince us that a succession of sounds is music, but the detection of any one of these trends allows us to infer that we may be in front of a musical event, so that we can then look for other trends that confirm this. Sometimes a sound organisation that is called music doesn't seem to have any particular trend on which to focus our perception; if this extreme situation is presented in a musical context (in a concert for example) it makes us look for organisation even if we cannot find it.

These simple trends and patterns give us some hints on how we could "teach" a machine to recognise that a sound phenomena is musical. From there on we should probably have to specialise the system in regards to certain styles and rules in order to obtain a dedicated analyst, capable of working on any period of time and detecting the style of the music. The next problem will be what particular search are we going to ask the system to perform in order to assist us or to uncover characteristics that are not evident through listening.

These simple conclusions lead us to think that being musical is something much simpler than we think; that there are more commonalities than differences to be found while listening. Another level appears here, concerning what we look for in music and how emotional states are possible through listening. I am getting out of the realm of artificial intelligence here to focus on human listening and the way pleasure or beauty are sought for and found. Experimentation and models are lacking here, so I will address this last point through introspection, even if it is a subjective and not reliable research source. I would then like to refer to my interests, and expectancies, when listening and making music.

5 My Personal Aims

Now, how does a musician live all this? Personally, I find music is one of the most fascinating domains to deal with. It is above all a pleasure domain in which I am capable — and eager I should say, of extracting knowledge from listening. I can experience high-level intellectual pleasure when listening to complexity and to structures, and at the same time feel a primitive pleasure when listening to some sounds or simple sound combinations. Pleasure is understood here at the emotional level generally related to aesthetic considerations (I like this sound, or what a beautiful melody), and at the rational level generally related to knowledge considerations (I understand what is happening, I perceive the complexity of a construction).

I am a listener, who reacts and analyses incoming musical information and I am a composer that builds sound structures for other listeners. As a composer, I have often analysed and questioned the influences of my listening in my work. The pleasure I obtain through music listening is the result of a long and still evolving accumulation of repeated experiences in which some patterns (melodic, harmonic, rhythmic or timbrical) are selected and memorised becoming strong regulators of expectancy. Pleasure comes through repetition of patterns built in my memory through time; they function as independent models that I may find in any music. The fact that some patterns become predominant has probably strong psychological implications and I am quite convinced that we actually look for those patterns while listening to any music.

It is also probable that I transfer in my creation some or all of these patterns and that finally composing, in my case, is building a scenario and a strategy for presenting and articulating these patterns. A strategy is deciding on which patterns I will work on and then building articulations between them or constructing interpolation. Patterns change and evolve, some are strong and have always been there, and others are weak and appear for certain periods. What I want to stress is that memory patterns are movable and subject to evolution, since musical listening is a constant and accumulative experience.

The differences between a system and myself are that I describe a system as being capable through self-learning to identify that a message is musical and then try to describe the functioning and the laws of the musical phenomenon I'm looking at. Through my perception, the path is quite similar; I look for musical organisation, when I decide it is musical, I compare it with previous experiences (memory) and classify it very quickly within a (generally) already known musical reference. However, my listening is oriented not towards the identification of known structures and rules (which I can also do), but to the repetition of memorised pleasure patterns that will strongly interfere with the way I orient my listening. These patterns can be found anywhere, in any music, and sometimes even in a non-musical context.

6 Let Us Keep on Searching

There is a lot to be found out there, and conferences like this one give us the opportunity to see the tendencies that move researchers to new propositions and results that confirm or contradict our thoughts.

High level modelling is one of the great issues for the next years, but as fast and effectively as we advance, the great difficulty is still in the transmission of high level concepts to signal processing levels. We do not know if linking the high and the low-level is the best solution, new models will appear that may allow us to simplify the comprehension of events. At the same time I am optimistic about our approach, I think there may emerge unexpectedly efficient methods that will simplify the tasks. Another solution could be a computer of such power that would permit us to do less algorithmic optimisation and would be capable of launching huge calculations, which infer results.

Two other great difficulties are associated with the task of understanding how music works; they are not the main objective of this presentation but they are important to mention since they may have strong influences on possible roads for the future.

The first difficulty concerns the making of meaning through time. Meaning depends on different time spans and it is built through the relation between our perception of the present and the memory of the immediate past that together permit an eventual projection towards the future. Joining the past and present to build the future is one of the main faculties we develop through listening, it implies widening the present so it can capture what has been and infer what may be. We have to be continuously looking back in time, in our memory; to look for organisation, regularity, homogeneity and try to build in our minds the plausible coherency, that will transform a bunch of sounds in music.

The second difficulty concerns the way we deal with sound in our minds. Sound is not only acoustic information that is dealt with as an independent media in our minds. Sounds create mental representations that we call images and are associated with the mental representations of the causes of the sound. When we listen to sounds we look for causes, real or virtual and we are quite good at imagining possible origins for all the sounds we listen to without seeing the causes (Acousmatics it is called). This is one of the great achievements of electroacoustic music, where the composer not only organises sound to create meaning but he can also manipulate images that will influence the way the listener perceives.

The creation of meaning and image representation are progressive experiences; the more I listen the more I accumulate memory patterns and the more I expand my capacities of perception through representations. We live in a century of repetition where we not only have access to any information but in the case of music, we have a repeated access to information. We can easily repeat the pleasure of listening once-again and access the same pleasure, be it intellectual, emotional or physical; and invent models capable of making our tools more effective and our understanding broader.

Extraction of Structural Patterns in Popular Melodies

Esben Skovenborg[1] and Jens Arnspang[2]

[1] Department of Computer Science, University of Aarhus
Åbogade 34,DK-8200 Århus, Denmark
Esben@skovenborg.dk

[2] Department of Software and Media Technology, Aalborg University Esbjerg
Niels Bohrs Vej 8, DK-6700 Esbjerg, Denmark
Arnspang@cs.aue.auc.dk

Abstract. A method is presented to extract musical features from melodic material. Various viewpoints are defined to focus on complementary aspects of the material. To model the melodic context, two measures of entropy are employed: A set of trained probabilistic models capture local structures via the information-theoretic notion of unpredictability, and an alternative entropy-measure based on adaptive coding is developed to reflect phrasing or motifs. A collection of popular music, in the form of MIDI-files, is analysed using the entropy-measures and techniques from pattern-recognition. To visualise the topology of the 'tune-space', a self-organising map is trained with the extracted feature-parameters, leading to the Tune Map.

Keywords: Melodic similarity, musical genre, feature extraction, entropy, self-organising feature map, popular music.

1 Introduction

Recently, the research area of music information retrieval (MIR) has received a lot of attention. Methods for organising and searching through collections of musical material are essential to MIR. Such skills are increasingly relevant, as more and more faith is attributed to the Internet as the future medium of music distribution. (Downie [6] provides an overview of the challenges and perspectives within MIR.) One type of MIR concerns *symbolic representations* of music. Music scores and Standard MIDI files are two common examples of such representations. Arguably, most music exists in sub-symbolic representations, like a recording of a performance in the form of an audio signal; but at least in principle the recording could be transcribed into some suitable symbolic form. (This task implies an exercise in abstraction, however, that is not yet possible without human assistance, even though polyphonic transcription is a classical research problem in music technology.)

Having a collection of music in symbolic form, one might want to search for a specific tune using a melodic query. In this case, when the melody is represented as a sequence of notes, the retrieval of the music is related to a string-matching problem [24, 10, 15].Moreover, the Multimedia Content Description Interface (or MPEG-7) provides a broad standard which includes definitions of certain features of a melody [18].

U.K. Wiil (Ed.): CMMR 2003, LNCS 2771, pp. 98–113, 2004.

One example of such an MPEG-7 Descriptor: *"the MelodyContourDS is a compact representation for monophonic melodic information, which facilitates efficient and robust melodic similarity matching. The MelodyContourDS uses a 5-level contour (representing the interval difference between adjacent notes)"* [17].

Now suppose that, instead of searching for a specific tune, we were interested in retrieving all the tunes in our collection that were *similar* to the given tune. Here, 'similar' ultimately means that we would subjectively judge the synthesis or performance of the tunes to sound similar (note that music in symbolic form, as such, is inaudible). The tunes that we would consider to be similar might be variations of the same melody, in which case MPEG-7's MelodyContour, or similar features computed for each tune, would presumably have let us predict the similarity. On the other hand, the tunes perceived as similar might be different melodies that were similar in *style*. In order to organise a collection of tunes according to stylistic similarity, we would need features that were able to capture one or more stylistic aspects of the material. Although the structural patterns model several musical characteristics, many aspects of musical style are clearly not dealt with; in general, this is partly due to limitations of the symbolic representation, and partly due to the overwhelming complexity of music perception.

The work presented in this paper introduces a method to extract high-level features which characterise structural aspects of the melodic material. By comparing these features of a collection of tunes, it is possible to organise the tunes according to a stylistic similarity; this corresponds to a cluster analysis within the feature space. In contrast, the string-match type of similarity-measures, which do not capture any deeper structures, would in general not provide a meaningful way to organise different melodies.

Musicology has a long tradition of (musicologists) formulating rules for the characteristics of certain musical genres and styles, in more or less rigid forms. But rather than explicitly defining rules that would characterise various stylistic properties, we employ a machine learning approach, such that an adaptive model is able to extract the relevant structures from a collection of existing melodic material. Thus, the model will 'learn' to recognise structural patterns in the material, simply because these structures are characteristic to the stylistic genre of the material.

In section 2 of this paper the symbolic representation is described, in particular which information in retained and which is discarded. Section 2.1 introduces the feature extraction method, and then the structural modelling based on entropy measures. In section 3, a collection of popular melodies is analysed and the result is visualised. Conclusions are presented in section 4.

The methods described in the following sections are based on an earlier project by this author [25].

2 Objectives and Melodic Representation

Popular music can be reduced to a melody-line with an accompanying chord progression. This is the type of musical material that is analysed and modelled in the present work. To allow us to analyse existing music collections, a method was developed to extract two note-sequences from a Standard MIDI File: *melody* and *harmony*. This pair of note-sequences is then quantised and transformed into a concise numerical representation, while retaining the most basic musical properties.

More specifically, the representation of melodic material used here, called the **Scale-NoteSequence**, has the following properties:

- concise symbolic representation which contains -
  - quantised melody and chords (i.e. a 'song book' level of detail)
  - no timbre or phrasing
  - no arrangement or form
  - no absolute pitch (notes are relative to a tonality)
  - no non-diatonic notes (only scale-notes)
- similar sounding melodic material $\leftrightarrow$ similar numerical representation
- automatic analysis and re-synthesis (via MIDI files)

In the case of the melody, the ScaleNoteSequence is simply a vector or string of notes, relative to the tune's diatonic scale. The advantage of a semantic value for each scale step (such as the *tonic*) was preferred to the ability to represent all notes (using the chromatic scale). The occasional non-diatonic notes are 'rounded' to the nearest scale-note. The notes are quantised at sixteenth-note resolution using techniques comparable to the quantisation functions of sequencer software like Cubase. For the harmony track, the quantisation level is quarter-note, and for each time step a cluster of notes is formed, to represent the most prevailing chord during that period. A cluster analysis of the tonal information, in the form of pitch (histogram) profiles, automatically produces the musical concepts of scale and tonality. This information is utilised to map from MIDI notes to scale notes and thus obtain a more musically relevant representation. Before each MIDI file is analysed, it is edited manually to ensure that the arrangement contains both a distinct melody- and harmony-track. Furthermore, the MIDI arrangement is 'cropped' such that the melody is reduced to one verse and one chorus (i.e., AB form). The ScaleNoteSequence is a skeletal representation which contains no information about the timbre or the phrasing of the melody-instrument, nor the 'swing' of the accompaniment. Nonetheless, for this project we shall make the following **assumptions**:

1. even when reduced to the ScaleNoteSequence representation, a melody retains some pertinent aspects of its musicality
2. this musicality is reflected in characteristic information structures
3. the specific structures vary according to the style/genre of the music

With the above assumptions, a collection of popular music as MIDI files, and a technique to transform it into ScaleNoteSequences, we can state the **objectives** for this work, being -

1. to model the information structures that are characteristic for the musical material
2. using adaptive modelling so that the specific patterns, characteristic for a genre, emerge from a training-set of music, rather than being explicitly defined
3. to extract high-level features (via the model) that can be used to define a stylistic distance measure

The developed distance measure may be used to compare and organise the music based on some of its stylistic properties, as motivated in the previous section.

2.1 Melodic Feature Extraction

At the surface level, music can be regarded merely as one or more parallel 'flat' sequences of tones. But *music listeners* have constructed a set of intuitions or expectations about the structure of music, as a consequence of their life-long experience with it. The listeners subconsciously impose their internal structures onto music they listen to, whereby musical meaning may be perceived from the sequences of tones.

In this work an adaptive model is presented to mirror and capture some of these intuitive 'grammars' of the experienced music listener. The underlying assumption is that the structures which are represented in the listener's internal model of music will be characteristic in the music itself. Principles of machine learning, techniques from nformation theory and statistical pattern recognition have been employed in our model to analyse the musical information and extract the characteristic patterns.

2.2 Viewpoints and Derivative Forms

Although the melodic material in the ScaleNoteSequence has been quantised in both pitch and time, it still represents a complex language, and would hence require a complex model and also a large data-set to learn from. So rather than attempting to model the melodic material directly, a family of viewpoints transform the ScaleNoteSequence into some simpler forms. The notion of music analysis by multiple viewpoints was coined by Conklin [4,5].

Each **viewpoint** will -

- define a transformation of the ScaleNoteSequence - possibly via a parent viewpoint - into a **derivative form**, focusing on a distinct aspect which is thought to be relevant to the perceived quality of the material
- discard information from the ScaleNoteSequence, thus leading to simpler structures in the derivative forms - easier to model

In total, 18 viewpoints are defined in three overlapping categories:

- **rhythmic** viewpoints, related to the structure of the ScaleNoteSequence, when the pitch information is discarded
- **tonal** viewpoints, related to pitch structure, when the rhythm information is discarded
- **hierarchical** viewpoints, related to the metrical structure

The 18 implemented viewpoints were essentially inspired from established phenomena in music theory and music psychology, see for instance [9,14,11]. Thus each viewpoint implements some domain knowledge about what structures are musically relevant, but does *not* impose rules to define what the (genre-)specific patterns are.

Many of the viewpoints belong in more than one of the above three categories, but they still discard some information; for example, one viewpoint is mapping the ScaleNote-values onto a binary chord-tone vs. other-scale-tone value. The harmony track, that was extracted together with the melody, is only used *indirectly*: it provides a key-centre that allows the MIDI-notes to be mapped onto ScaleNotes, and is additionally used by some of the tonal viewpoints to determine whether each ScaleNote

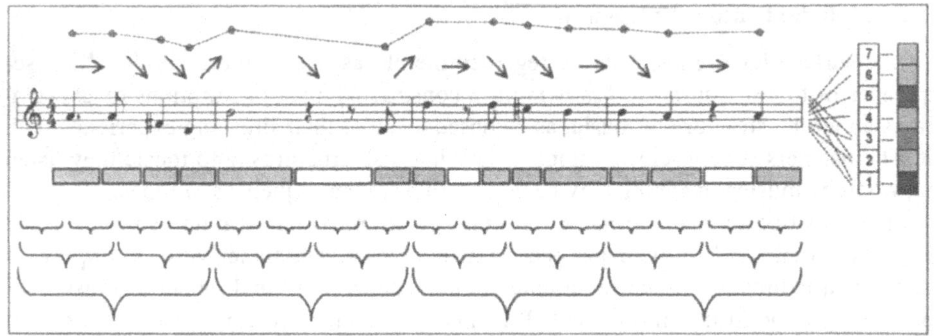

Fig. 1. Illustrates the different types of viewpoints: tonal, rhythmic and hierarchical

is a chord-tone. Thus, only the melody information is exploited by the present model, although it might be advantageous, in future work, to model the underlying chord progression as well.

A number of the derivative forms of each viewpoint type are illustrated in Figure 1, for a four-measure melody sample. The ScaleNoteSequence is depicted in standard music notation, as the two are nearly equivalent melody representations. It is beyond the scope of this paper to define and justify each individual viewpoint. But, for example, the DeltaScaleNoteOn form represents each adjacent pair of pitches (in the ScaleNoteSequence) by the difference between them, counted in scale-steps. This derivative form does not represent any rhythmic, absolute pitch, nor metrical information; it is thus an example of a *pure tonal* viewpoint. The arrows above the score in Figure 1 illustrate the MelContour viewpoint, which can be regarded as a tertiary version of the MPEG-7 descriptor in the Introduction section.

The horizontal sequence of grey-shaded rectangles in the middle of the figure illustrate several *pure rhythmic* viewpoints: one of these simply transforms the ScaleNoteSequence into a sequence of note- and rest-durations. The braces, in the bottom of the figure, illustrate the *hierarchical* type of viewpoints, focusing on metrical information. For instance, one such derivative form discards all the notes not on 'strong beats'. The numbers 1-7, to the right of the score, represent the scale-degrees: in the example the note D = *scale degree 1*, and so forth.

The above figure shows how focusing on individual aspects of the melodic material will lead to simpler information-streams, easier to model than all properties at once (the parent ScaleNoteSequence).

2.3 Markov Models and Entropy

Replacing the original ScaleNoteSequence, the viewpoints provide 18 different derivative forms which focus on musically relevant aspects of the melodic material.

Now let's return to the task of comparing two different melodies. Our viewpoint-analysis would for each melody provide a set of 18 symbol-sequences, which would leave us with the basic choice of either performing string-match sort of operations on the pairs of derivative forms, or somehow *characterising* the information in each sequence so that it can be quantified and compared as a whole. The latter approach is pursued here, as we are after a stylistic comparison rather than a phrase-matching. One way of characterising the derivative forms would be to compute and compare the statistics of the sequences, like the 'mean note-duration' or the 'pitch-interval spread'. However, instead of measuring such surface-features, we use a *model-based approach* in which the models can learn the characteristic patterns of a given style, and thus also capture temporal structure.

With his pioneering work, Claude Shannon [22] presented a novel quantifiable measure of information content - its unpredictability or *entropy*. He describes how we may consider sequences of symbols (messages), from a finite alphabet, to be generated by a discrete source of information, according to probabilities which depend on the preceding choices of symbols. The definition of the entropy measure is based on these probabilities. Natural written languages are used as examples of such stochastic processes [23].

Prior to calculating the entropy of messages, it is necessary to construct a stochastic model of the information source in order to supply the probabilities of the possible events. We clearly cannot accept the assumption that symbols in the derivative forms are independent on the previous context. Therefore we must consider the *conditional probabilities* of the events. Finite-context models provide more accurate entropy measures than the 'zero-memory model', when symbols are dependent on the previous context. An *mth* order *Markov model* contains probabilities which model the dependency of the next symbol on the preceding *m* symbols in the sequence [3].

We construct and train (estimate transition probabilities for) a Markov model corresponding to each of the implemented viewpoints. High model-order may lead to more precise estimates of entropies for new messages. But, as the number of model states depend exponentially on the order, the *curse of dimensionality* imposes practical limitations regarding the size of the transition-matrix. Model orders of 2-4 are used, depending on the number of states for each model, which again depends on the size of the alphabet used by the corresponding viewpoint.

Once trained, each Markov model can estimate the entropy of any new derivative form produced by the same viewpoint. Thereby we obtain a relevant quantifiable measure of the 'musical (dis)order' that we can use to compare two sequences, and contribute to a measure of their distance.

For popular music tunes, the entropy of the 'raw' ScaleNoteSequence form - which contains enough information to re-synthesize the analysed melody — is about 1.5 bits/symbol. This is equal to 3 bits per eighth-note duration of melody, which is interesting because the estimated entropy of written English is around 3 bits/letter [3].

Applications of information theory to music started in the early 60'es, although experiments were limited by the available computational resources. However, the standpoint that music could basically be described as a system of probability relationships, the complexity and form of which would depend on the style, has been criticised for being simplistic and inappropriate — and rightfully so[1].

[1] A somewhat similar situation applied to classical AI, in which the music was modelled instead by rule-based systems.

According to Krumhansl [14], *"Information theory rapidly extended beyond the engineering applications for which it was originally developed. It had considerable impact on psychology in particular, in which it was applied to both linguistic and perceptual information. [...] Information theory also had some influence in the area of music, in which music was compared to a transmitted message. In this context, the question was whether it would be possible to ascertain the level of uncertainty that is optimal. Music that is too simple, with too much redundancy, is perceived as uninteresting; however, very complex or random music is perceived as incomprehensible. [...] Another question asked in this context was whether the measure of information would distinguish between different musical styles or different composers."*

More recently, similar ideas have been pursued with greater subtlety, see [20, 5, 26]. The latter paper describes a computational model to predict melody-notes of Bach chorales. The model, maintaining both a long-term knowledge base and short-term 'transitory constraints', was trained on different musical aspects of a sample set of chorales, and its predictions result in 'entropy profiles'. Human listeners were given the task of guessing the notes of certain chorale-melodies, and the results can be interpreted as entropy profiles as well. This facilitates a quantitative comparison of the performance of the computational model with that of the human listener. Witten and colleagues report that the entropy-profiles are similar for the two 'systems', and in particular that they encountered difficulties at the same points in the chorales, indicating *"...that it is fundamental musical characteristics that are being measured [by the entropy]"*. The humans still outperformed the constructed model, but it is suggested that *"...it does not seem unlikely that machines will eventually outclass people in predicting a restricted and well-defined genre of music."*

2.4 Motifs and LZ-Coding

Throughout a composition certain short melodic or rhythmic patterns are typically repeated, and there are many ways in which the composer can modify each instance of the pattern, depending on the genre, his style, and the musical context. When such a recurring musical pattern is recognised by the listener it is known as a theme or a *motif*. As the listener identifies musical sequences as variations, expectations are generated, which the composer can either fulfil, or (ab)use to produce musical surprises.

Pertinent use of motifs is an essential aspect of melodic material, and several of the derivative forms were intended to highlight tonal as well as rhythmic motifs, in order to quantify these aspects. However, the Markov models do *not* capture such recurring patterns — once the models are trained, they cannot adapt in any way to the individual sequence: a repetition of some phrase is not distinguished from any other phrase with the same entropy. Therefore another type of processing, based on *adaptive encoding*, is introduced to supplement the Markov models.

A technique of adaptive dictionary encoding, named *Lempel-Ziv coding*, is based on replacing repeated phrases in a sequence with a pointer to where they have occurred earlier. Each element in the encoded output is called an *LZ-piece*. The original symbol-sequence can be reconstructed left-to-right from the LZ-pieces output by the encoding algorithm. Although no dictionary or probabilities are coded explicitly, the encoding of any symbol sequence containing regularities in the form of repeated phrases will

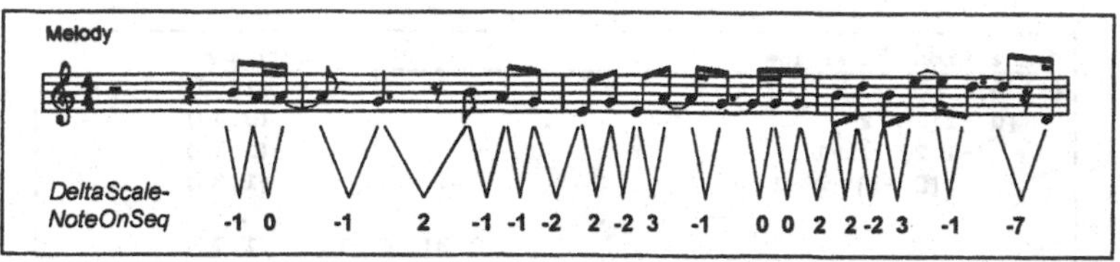

Fig. 2. The derivative form DeltaScaleNoteOn is shown in this example with 4 measures of the tune Desperado(The Eagles, 1973)

typically be shorter than the sequence itself - hence the algorithm's called a 'universal compressor'.

LZ-coding comprises a family of algorithms, generally derived from one of two different approaches published by Abraham Lempel and Jacob Ziv in 1977 and 1978, labelled LZ77 and LZ78 respectively [3]. The algorithm implemented here is a hybrid between the LZSS and the LZR algorithms (both of which are based on LZ77), as this leads to the combination of properties most suitable for the task at hand.

As a complementary feature of certain derivative forms, we use the structure of the LZ-pieces resulting from the LZ-encoding of the sequence. On these structural properties we base our quantitative measure of *to what degree phrases are repeated in a sequence*. Thereby, the set of trained Markov models can be said to capture *local* structure of a derivative form, while the LZ-encoding captures its *global* structure.

2.5 An Example Viewpoint

In this section, the tonal viewpoint *DeltaScaleNoteOn* is used as an example of the techniques described in the preceding sections. Figure 2 shows the corresponding derivative form, calculated for with the first four measures of a well-known popular melody. The example melody is shown in common music notation, representing (the information of) the parent ScaleNoteSequence. The numbers underneath the melody are the DeltaScaleNoteOn derivative form, which is calculated by subtracting each scale-note value by the previous note's value: *DeltaScaleNoteOnSeq(i) = ScaleNoteOn(i+1) - ScaleNoteOn(i)*. Note that all rhythmic and metrical information is discarded by this tonal viewpoint.

The two different measures of entropy, introduced in section 2.3 and 2.4, are demonstrated in Figure 3 and Figure 4, respectively. The same four measures of the example melody are used again.

The Markov model, demonstrated in Figure 3, characterises local structures in the derivative form; the context length is two symbols because a 2. order model is used. Each symbol in the sequence, together with its context, corresponds to a transition in the model. The entropy value for each transition is based on the probabilities provided by the Markov model. The model's probability transition matrix is estimated during the foregoing training-phase, based on a collection of popular music.

The example of the Markov model entropy shows how a DeltaScaleNoteOn of -7 with the context [3,-1] has the relatively high entropy of 5.21 bits. This means that this

```
Transition / message                               Entropy
[-1 0] → -1                                         (1.68)
   [0 -1] → 2                                       (3.49)
     [-1 2] → -1                                    (1.53)
        [2 -1] → -1                                 (1.38)
                          . . .                      . . .
                                  [-2 3] → -1       (1.78)
                                     [3 -1] → -7    (5.21)
```

→ ***Example parameter:*** Mean entropy = 2.496

Fig. 3. The trained 2. order Markov model, corresponding to the example viewpoint, calculates a measure of entropy based on the local transitions in the derivative form's symbol sequence

-1 | 0 | -1 | 2 | -1 | -1 | -2 | 2 | -2 | 3 | -1 0 | 0 | 2 | 2 -2 3 -1 | -7 | ...

→ ***Example parameter:*** Longest LZ-piece = 4

Fig. 4. The adaptive coding of the example derivative form by the LZ-algorithm. The first time a new symbol appears in the sequence, it is encoded as a Symbol (underlined). In all other situations one or more symbols, already encoded, can be matched and hence constitute an LZ-Piece, separated by bars in the figure (the actual pointers to the previous locations are omitted from the figure for clarity)

particular transition - the pitch descending an octave after having gone up three scale-notes and down one — is quite unlikely in the style on which the model is trained. This way, deviations from what the model has learned are quantified for all the tonal, rhythmic and hierarchical viewpoints. If two different tunes would deviate in the same manner, this would be reflected in their respective feature-vectors, and they would consequently be considered (more) similar.

The Lempel-Ziv algorithm's encoding of the same symbol sequence is shown in Figure 4. In contrast to the Markov model, the LZ-algorithm requires no training, and therefore has no /'a priori expectations about the sequence. The example demonstrates how the algorithm detects recurring sub-strings in the sequence, using the entire preceding part of the sequence as context - hence the term *adaptive encoding*.

Note that the viewpoint shown in the example (Figure 2) is based on the *change* in scale-note (the 'delta'), and it is therefore unaffected by the actual scale-note values. Thus the LZ-algorithm detects the recurring 4-note motif in the example even though it is transposed the second time it occurs. This motif has a DeltaScaleNoteOn sub-sequence of [2,-2,3,-1]. The longest LZ-Piece is 4, for the example, which would distinguish this melody from another melody with similar local structures (or surface features), but without the tonal motifs.

2.6 Summary of the Melodic Feature Extraction

The diagram in Figure 5 shows the overall data flow in the model presented, from the music in MIDI file form, to the final feature-parameters.

Currently, **the MIDI files must fulfil certain conditions:** For correct extraction of the melody-track leading to the ScaleNoteSequence, the melody must reside on its own track inside the MIDI file and must be (close to) monophonic. Similarly, the extraction of the harmony requires the chords to appear as 'clusters' on their own MIDI track, e.g. as opposed to arpeggio-style. Furthermore, the material must have a 4/4 meter, and the melody must fit within a sixteenth-note quantisation. These limitations could, in principle, be overcome by extending the system's parser of the MIDI note data. Only a **few general assumptions were made about the music** to be modelled. The most severe assumptions are that the melody should fit on a diatonic scale, and the chord progression (harmony) should provide a stable key-centre. In practice, these assumptions are not very limiting when the material in question is popular music. For instance, the melodies rarely contain non-diatonic notes, and mainly as 'leading tones'. Also, the key-centre requirement easily leaves room for, for instance, a change to a related key between the verse and chorus. Within the framework of the assumptions and the implemented viewpoints, the stylistic rules that the model learns depend entirely on the material that it is trained on. Two different techniques from the field of information theory are applied to characterise the patterns of the melodic material on a local and on a global scale, respectively.

The so-called **parameterisation of the entropy-estimates**, produced by the Markov models and the LZ-encodings, is done using standard techniques from statistical pattern recognition, like measures of the central tendency and spread [7]. Thus a fixed-length vector of parameters may be computed for melodic material of any length.

Because the raw parameters are based on derivative forms that are overlapping in the information they represent, the parameters are correlated. A *principal component analysis* is applied to the parameters with two benefits: the resulting feature-parameters have a lower dimensionality, and they are de-correlated [12]. The dimensionality can be reduced from 150 to 18, while retaining 85% of the variance in the feature parameters.

The feature-vectors resulting from the analyses of different input tunes, may be compared using a *standardised Euclidean distance* measure. This facilitates a quantitative comparison of melodic material, based on structural patterns in the melodies, and this constitutes our proposed stylistic distance measure.

3 Clustering and Visualisation

The presented model was trained on a data set consisting of popular music. To visualise the structure of the feature-space and the resulting relationship between the tunes, a self-organising feature map was employed.

3.1 The Training Set

A collection of 50 tunes constitute a sample from the established popular music. Tunes in this musical genre normally have a strong reliance on the melody, which makes them well suited for analysis by the presented method. Standard MIDI-files for these tunes, which comply with the format assumptions of the system, have been produced. The form of each tune is either verse-chorus or simply verse, and the MIDI files have been edited to include only one time through this basic form.

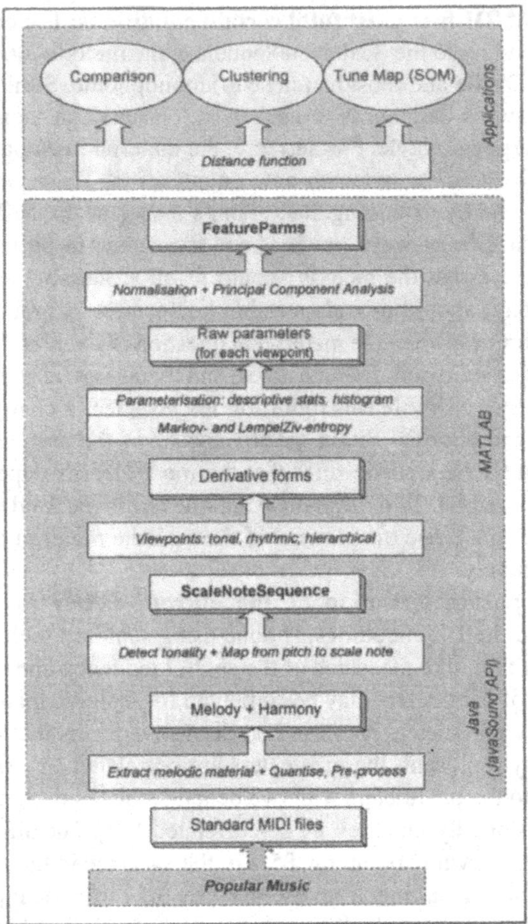

Fig. 5. Overview of the analysis and feature extraction

The tunes in the collection belong to two different musical styles:

- **33 contemporary international pop-songs**. Composed during the past three decades, most of these songs are known and recognised by the average music listener - they are hits and have become classics.
- **17 Danish traditional songs**. Most of these tunes were composed at least one century ago, and in many cases knowledge of their origin and composer is lost [16]. They have survived until present time mainly by virtue of the memorable quality of their melodies (and lyrics).

Tunes from both stylistic categories of popular music have in common that they are characteristic and simple enough to be remembered by musically untrained listeners. Yet each tune is easily classified (by the same listeners) according to its style, even when re-synthesized from a ScaleNoteSequence level-of-detail. These qualities qualify the selected tunes to constitute the data-set for training the Markov models and thereby the principal components, i.e. setting the free parameters of our feature extraction model.

3.2 The Tune Map

A Self-Organising feature Map (SOM) is often considered a kind of neural network. The SOM consists of *nodes* which are arranged geometrically according to a topology function, usually in a two-dimensional grid. Each node is associated with a *reference vector*, corresponding to a point in the input space. During training, the reference vectors are fitted to the samples of the training data set; but for each sample both the best-matching node as well as its neighbours are updated. Thereby the SOM training-algorithm tries to organise the feature-map such that any two nodes that are close to each other in the grid have reference vectors close to each other in the input space. Thus the trained SOM can position input vectors, in the grid, according to their clustering in the input space. Further details, including training algorithms, can be found in [13]. The functions in this work, for training and visualising the SOM, were based on the 'SOM Toolbox' [1].

The SOM can be regarded a multi-dimensional scaling method which projects the data from the input space to a lower-dimensional output space (the grid). Another common explanation is that a SOM tries to embed an 'elastic' grid in the input space, such that every training sample is close to some suitable reference vector, but the grid is bent or stretched as little as possible. In effect, the self-organising map learns both the distribution and topology of the input vectors of the training data set. Note that the SOM (in the form used here) is a case of *unsupervised learning*, as no target vectors or class information is part of the training set.

Its capability of visualising a high-dimensional vector space, and discovering both clustering and cluster-relationships by unsupervised learning, makes the Self-Organising feature Map suitable for investigation of the 'tune-space' constructed by our melody feature extractor. Therefore a SOM is trained on the feature-vectors resulting from the model's melody analysis, thus producing a *Tune Map*. Figure 6 shows the Tune Map of the tune collection described in section 3.1. This collection is composed of popular tunes from two different eras, and it is quite conceivable that the patterns of the musical idiom for one group of popular melodies would be somewhat different from the patterns of another group of popular melodies, composed over a century earlier. This *á priori* classification of the tunes, as either contemporary international pop or Danish traditionals, is shown on the SOM above, by printing tune-labels of the latter group in inverse. It is unavoidable to observe that these two groups of tunes have been separated in the trained SOM.

As all the tunes of the collection have been processed identically (except for their labels), the observed grouping of the data in the SOM must reflect a similar organisation in the feature-space. In other words, the extracted feature-parameters capture certain properties of the melodic material which characterise its musical idiom sufficiently to distinguish Danish traditionals from contemporary pop.

It would not be unreasonable to expect that other musical properties, represented in the features, are reflected in the topology of the SOM as well. However one must avoid the temptation to over-interpret the results, such as the fact that both of Denmark's

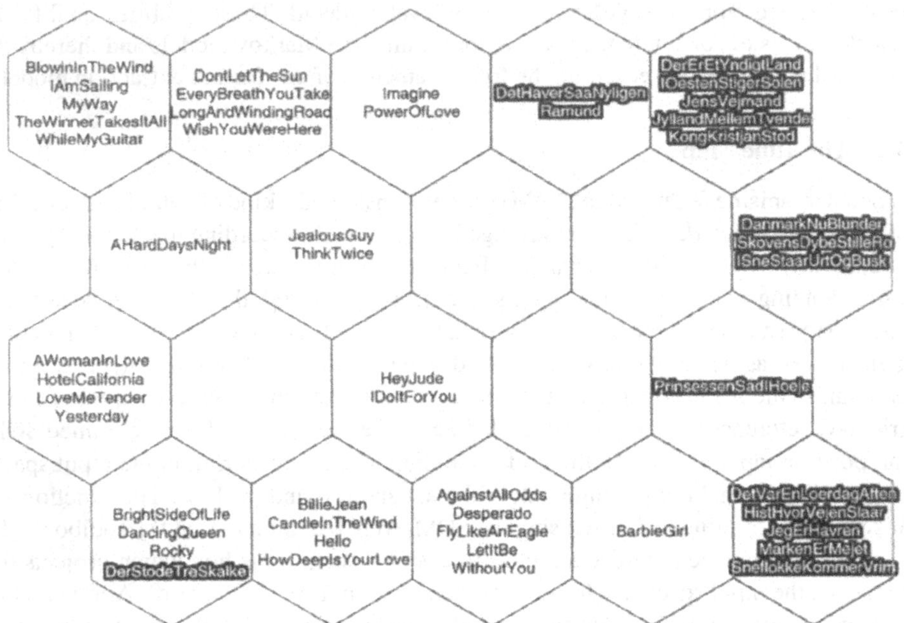

Fig. 6. The Tune Map of 50 popular melodies from two different eras: Danish traditional songs with inverse titles, and contemporary international pop songs with normal print titles

national anthems share the very same node (top right). A closer investigation of the trained SOM involves inspection of its component planes and reference vectors, but is beyond the scope of this paper.

Projections of a high-dimensional feature space onto a two-dimensional SOM may lead to a rather 'convoluted' map. As an interesting alternative to increasing the dimensionality of the SOM (making it harder to visualise), a hierarchical structure of SOMs can be used [21].

4 Conclusion

In this paper, we have presented a method to characterise melodic material in symbolic form. Features are extracted by which a measure of similarity of different melodies can be computed. The features are constructed from a set of viewpoints, to focus on complementary tonal, rhythmic, and metrical aspects of the melodic material. Inspired from information-theory, two measures of entropy are employed, both to capture the unpredictability on a local scale, and to capture phrasing or motifs on a tune-global scale. By methods of statistical pattern recognition, the entropy-measures are transformed into the feature-parameters that can be applied to a stylistic comparison of melodic material.

Unsupervised learning is employed at several stages in the model: The detection of tonality is based (only) on pitch-profiles in the training data; also optimised in the training-phase are the probabilities inside the 18 Markov models characterising the derivative forms, and the projection of the PCA leading to the final feature vector. In this study, the model was trained on a small collection of popular tunes, and the free parameters of the overall model were thus optimised towards this genre. But in principle, this music collection could be substituted with a set of melodic material from a different genre, and by repeating the unsupervised-learning steps the model could simply be re-trained. It must be assumed, however, that the realised viewpoints still focus on the relevant properties of the new material. A new genre, in which pitch, rhythm, tonality and metrical structures were *not* the significant properties, would require re-formulation of the viewpoints.

The system was tested using a collection of popular music tunes of two different eras. A self-organising feature map, trained on the full set of feature-vectors, demonstrated an unsupervised clustering of the tunes in the collection - the Tune Map. This resultant organisation clearly reflected that two different stylistic genres were represented in the collection. Thus the feature-parameters have evidently captured some properties of the melodies which characterise their individual musical idiom sufficiently to distinguish Danish traditionals from contemporary international pop. Note that the self-organising feature map is *not* a classifier, trained to discriminate between traditional and modern popular music; the system was never told that the music collection consisted of two groups of material. The fact that the Tune Map reflects this stylistic grouping means that a corresponding clustering naturally takes place along certain feature dimensions.

But what (other) high-level characteristics of the melodies does the organisation by the feature map correspond to? And which musical attributes are really captured by the developed similarity measure? One problem, in answering these questions, is the general difficulty of 'extracting knowledge' from statistical pattern recognition models. But perhaps the main obstacle, in the interpretation[2] of the developed features in terms of musical stylistic properties, might be that no objective and rigorously defined high-level musical features, on which to base the interpretation, are currently available! If musicology could provide, for instance, a theory to systematically categorise musical material, we could simply train a classifier on the categories, and would then be able to inspect which features were useful for which discriminations, and also test the generalisation ability of the system. Alas, such musical knowledge is not yet available, although the issues are topics of active research. Several existing taxonomies of musical genre, used in the music industry, have been studied, and their inadequacies and inconsistencies demonstrated [19].

The presented melodic similarity measure could possibly find applications within the MIR community. Note that the developed melodic features could be used either in a supervised learning paradigm, such as classification into known genres; or in unsupervised organisation, as the clustering demonstrated in this paper.

[2] In the foregoing project (Skovenborg, 2000), a genetic algorithm was designed to evolve populations of novel melodic material, with a fitness function based on the same kind of similarity-measure as presented here. Via MIDI-files, synthesized from the 'fittest individuals', the quality of the evolved melodies – and thus the melodic features – could be assessed by listening. Unfortunately, due to the evolutionary optimisation procedure, this method of algorithmic composition is rather time consuming. (This work is to be described in an upcoming paper.)

Systems for content based retrieval have been developed for querying a music database by presenting a melodic search-phrase. With melodic features characterising the style of the material, it might be possible to move beyond the string-matching types of query to a *stylistic query* in a collection of music.

Acknowledgements

Thanks to Søren H. Nielsen for his constructive critique of the manuscript. We would also like to thank the anonymous reviewers of this paper for providing insightful observations and suggestions.

References

1. Alhoniemi, E., Himberg, J., Kiviluoto, K., Parviainen, J. & Vesanto, J. (1997) SOM Toolbox Version 1.0. Laboratory of Information and Computer Science, Helsinki University of Technology. Internet WWW page at URL: <http://www.cis.hut.fi/>.
2. Aucouturier, J.-J. & Pachet, F. (2002) Music Similarity Measures: What's the Use? In Proc. of the Third International Conf. on Music Information Retrieval (ISMIR-2002), pp.157-163.
3. Bell, T.C., Cleary, J.G. & Witten, I.H. (1990) Text Compression. Englewood Cliffs, NJ: Prentice Hall.
4. Conklin, D. & Cleary, J.G. (1988) Modelling and generating music using multiple viewpoints, In Proc. of the First Workshop on AI and Music (AAAI-88), pp.125-137.
5. Conklin, D. & Witten, I.H. (1995) Multiple viewpoint systems for music prediction, Journal of New Music Research, vol.24:1, pp.51-73.
6. Downie, J.S. (2003) Music Information Retrieval. Ch. 7 in Cronin, B. (ed.) Annual Review of Information Science and Technology 37, pp.295-340. Information Today.
7. Duda, R.O., Hart, P.E. & Stork, D.G. (2000) Pattern Classification (2.ed.), New York: John Wiley & Sons.
8. Foote, J. (1999) An overview of audio information retrieval, Multimedia Systems, vol.7:1, pp.2-10.
9. Handel, S. (1989) Listening: an introduction to the perception of auditory events. Cambridge, MA: The MIT Press.
10. Hewlett, W.B. & Selfridge-Field, E. (eds.) (1998) Melodic Similarity - Concepts, Procedures and Applications. Computing in Musicology, vol.11. Cambridge, MA: MIT Press.
11. Howell, P., West, R. & Cross, I. (eds.) (1991) Representing Musical Structure, London: Academic Press, Cognitive Science Series.
12. Jolliffe, I.T. (1986) Principal Component Analysis, New York: Springer-Verlag.
13. Kohonen, T. (1997) Self-Organizing Maps (2.ed.), Berlin: Springer-Verlag.
14. Krumhansl, C. L. (1990) Cognitive Foundations of Musical Pitch, Cambridge, UK: Cambridge University Press.
15. Lemström, K. (2000) String Matching Techniques for Music Retrieval, PhD Thesis, Dept. of Computer Science, Report A-2000-04, University of Helsinki.
16. Madsen, L.B., Grøn, J. & Krøgholt, D. (eds.) (1997) Folkehøjskolens Sangbog (17.ed.), Gylling, Denmark: Foreningen for folkehøjskolers forlag.
17. MPEG (2001) Information Technology - Multimedia Content Description Interface - Part 4: Audio (ISO/IEC CD 15938-4, part of MPEG-7), ISO/IEC JTC 1/SC 29/WG 11.
18. MPEG Requirements Group (1998) MPEG-7 Context and Objectives (ISO/IEC JTC1/SC29/WG11, N2460), International Organisation for Standardisation.

19. Pachet, F. & Cazaly, D. (2000) A Taxonomy of Musical Genres, In Proc. of Content-Based Multimedia Information Access Conference (RIAO), Paris.
20. Ponsford, D., Wiggins, G. & Mellish, C. (1999) Statistical learning of Harmonic Movement. Journal of New Music Research, vol.28:2.
21. Rauber, A., Pampalk, E. & Merkl, D. (2002) Using psycho-acoustic models and SOMs to create a hierarchical structuring of music by sound similarity. In Proc. Int. Symposium on Music Information Retrieval (ISMIR), Paris.
22. Shannon, C.E. (1948) A mathematical theory of communication. The Bell System Technical Journal, vol.27, pp.379-423, 623-656.
23. Shannon, C.E. (1950) Prediction and Entropy of Printed English. In Sloane, N.J.A. & Wyner, A.D. (eds.) (1993) Claude Elwood Shannon: Collected Papers. New York: IEEE Press.
24. Skovenborg, E. (1997) Musik Database. Unpubl. project report, Computer Science Dept., University of Copenhagen. (Reviewed in Barchager, H. (1997) Datalogisk løsning pået musik pædagogisk problem. Anvendt Viden, vol.97:3, Videnskabsbutikken, Københavns Universitet).
25. Skovenborg, E. (2000) Classification and Evolution of Music - Extracting Musical Features from Melodic Material, Master's Thesis, Computer Science Dept. (DIKU), The University of Copenhagen.
26. Witten, I.H., Manzara, L.C. & Conklin, D. (1994) Comparing Human and Computational Models of Music Prediction. Computer Music Journal, vol.18:1, pp.70-80.

Musical Pattern Extraction Using Genetic Algorithms

Carlos Grilo[1,2] and Amilcar Cardoso[2]

[1] Departamento de Engenharia Informática da Escola Superior Tecnlogia e Gestã de Leria
Morro do Lena, Alto Vieiro, 2401-951- Leiria, Portugal
[2] Centro de Informática e Sistemas da Universidade de Coimbra
Polo II, Pinhal de Marrocos, 3030 - Coimbra, Portugal
{grilo,amilcar}@dei.uc.pt

Abstract. This paper describes a research work in which we study the possibility of applying genetic algorithms to the extraction of musical patterns in monophonic musical pieces. Each individual in the population represents a possible segmentation of the piece being analysed. The goal is to find a segmentation that allows the identification of the most significant patterns of the piece. In order to calculate an individual's fitness, all its segments are compared among each other. The bigger the area occupied by similar segments the better the quality of the segmentation.

1 Introduction

In artificial intelligence, it is common to name the process of identifying the most meaningful patterns occurring in some piece of data as *pattern extraction*. An important branch of this investigation area is dedicated to the problem of identifying the most meaningful patterns in data that can be represented as strings of symbols. This is a problem of great relevance in areas like molecular biology, finance or music. In the particular case of music, the identification of these patterns is crucial for tasks like, for example, analysis, interactive on-line composition or music retrieval [1].

In previous work done in this area ([2–4]) the musical piece to be analysed can be almost exhaustively scanned, so that all the existing patterns are identified. After that, some criteria are applied so that the most meaningful patterns can be identified. While this may be an effective procedure, we think that it is reasonable to investigate the possibility of identifying the most meaningful patterns existing in one musical piece without searching the entire space of its segments.

This paper describes a research work in which we study the application of genetic algorithms to the extraction of musical patterns in monophonic musical pieces. The two main reasons for choosing genetic algorithms to this type of problems are: the search capacity already demonstrated by these algorithms in very complex problems; the possibility of representing individuals as possible segmentations of the piece being analysed. This way, if we guide the search such that segmentations with the most meaningful patterns are favoured, at the end it will not be necessary to do extra processing. Since until now we have been more concerned with the question of "how can this be done?" and with "does it solve the problem?" than with "how fast it is", this paper will not cover aspects related to time performance.

U.K. Wiil (Ed.): CMMR 2003, LNCS 2771, pp. 114–123, 2004.

We start, in Section 2 by describing the prototype system we developed, named EMPE (Evolutionary Musical Pattern Extractor). In Section 3 we show the results obtained in some experiments that we have done. Finally, in Section 4 we draw some conclusions and indicate some future steps in our research.

2 EMPE Description

EMPE uses a genetic algorithm to discover the most significant patterns existing in a musical piece. It receives the piece to be analysed as an argument and returns a segmentation of that piece.

2.1 The Genetic Algorithm

Genetic algorithms are parallel and stochastic search algorithms inspired by the evolution theory and molecular biology, which allow the evolution of a set of potential solutions to a problem. In our approach, we use is a typical genetic algorithm, although with some changes. It starts by randomly creating an initial population *P(0) of size n* of potential solutions to the problem (each one represents a segmentation of the piece being analysed; see Section 2.2). Then, it evaluates this population using a fitness function that returns a value for each individual indicating its quality as a solution to the problem.

After these two steps it proceeds iteratively while a stop condition is not met. In each iteration, the first step consists of the creation of a new population *P1(t)*, also of size n, by stochastically selecting the best individuals from *P(t)*, the current population. Among the several existing selection methods, we use the tournament selection method [5], with tournament size of 5. After the selection step, a new population *P2(t)* is created by stochastically applying genetic operators to the individuals of *P1(t)*. The genetic operators we use are *classical two-point crossover and classical mutation* [5]. After the application of genetic operators we introduce another step that, while not being a new one, doesn't make part of typical genetic algorithms. It consists of the application of another type of operators, usually called learning operators [6], to some individuals of *P2(t)*. This step is inspired by the fact that the performance of biological organisms can be improved during their lifetime through learning processes. In artificial systems, some common differences between genetic and learning operators are: genetic operators have a predefined probability of being applied to any individual, while learning operators are usually applied only to a small randomly chosen set of individuals; learning operators, differently from genetic operators, are usually deterministic and are sequentially and repeatedly applied to each chosen individual during a fixed number of iterations or until their application stops improving the individual's fitness. The learning operators used in our algorithm will be described in section 2.5.

In the last step of the cycle, the modified *P2(t)* becomes *P(t+1)*, whose individuals are then evaluated using the already mentioned fitness function. Finally, after the search process is terminated the learning operators are applied to the best individual found, so that some last improvements can be made.

2.2 Individual's Representation

In EMPE, each individual stands for a segmentation of the piece being analysed, i.e., it consists on a sequence of segments in which the piece can be divided. 1 illustrates how individuals are represented and how they must be interpreted.

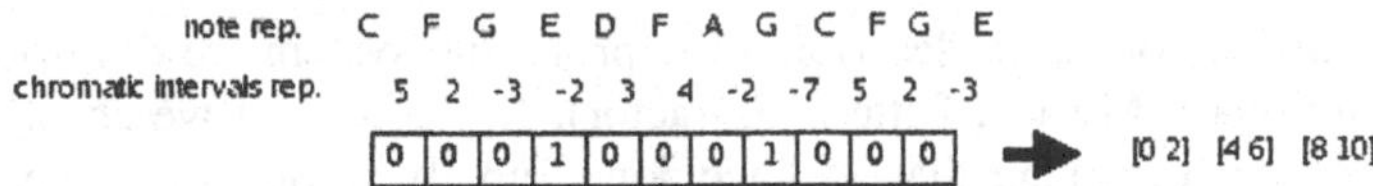

Fig. 1. A sequence of chromatic intervals and an individual representing a segmentation of it

As we can see in this figure, each individual consists of a binary string. All individuals of the population have the same length, equal to the number of intervals of the piece to be analysed (although in 2 we only show information relative to pitch, we actually represent musical pieces as sequences of tuples *chromatic_interval, duration_ratio*). The *i-th* digit of the binary string corresponds to the *i-th* interval of the piece. Segments simply correspond to "0" sequences, being the "1" s to represent intervals that, being between segments, do not belong to any. Following this interpretation, the individual of Fig. 1 corresponds to the interval segments sequence "5, 2, -3, 3, 4, -2 5, 2, -3", equivalent to the note segments sequence "C, F, G, E, D, F, A, G, C, F, G, E".

Due to practical reasons, before the learning operators application step, we convert individuals into a representation of the type "$[l_o\ r_o]$ ... $[l_n\ r_n]$", where $[l_i\ r_i]$ stands for the *i-th* segment of the piece, and l_i and r_i represent, respectively, the left and right limits of that segment. This type of representation, besides being a more understandable one, has the advantage of allowing an easier access to segment limits, which among other things, facilitates the evaluation process. Thus, in practice we have two types of representation: one we refer to as the genotype, on which genetic operators are applied, and another we refer to as the phenotype, into which individuals are converted before the learning operators application step. In order to facilitate the reading, in the rest of the paper we will use preferably the phenotype representation instead of the genotype one.

In EMPE, as we allow the user to define upper and lower limits to segments' length through, respectively, parameters *min_seg_len* and *max_seg_len*, it may happen that the application of genetic operators generates individuals with segments that violate these limits. In order to correct these situations, all individuals resulting from the application of genetic operators are submitted to the following correction procedure: if a too small segment occurs, concatenate it with the next segment (this is done easily by changing the "1" between the two segments by a "0"); if a too big segment occurs, divide it in two equal length segments.

2.3 Fitness Evaluation

Individuals, as described in the previous section, do not say anything about which segments are patterns. This information is attached to each individual during the evaluation process in the form of relations between segments, as we will now see. The fitness of an individual is calculated by comparing all its segments among each other. If the distance between two segments is less than a given value, they are considered similar and a relation between them is created. If this is not the case, no relation is created and the two segments are not considered as similar. Every time a relation is created, the difference value between the two segments is attached to it. In Fig. 2 we show an individual and the relations that are created as a result of this process (in this case, just one). Each rela-

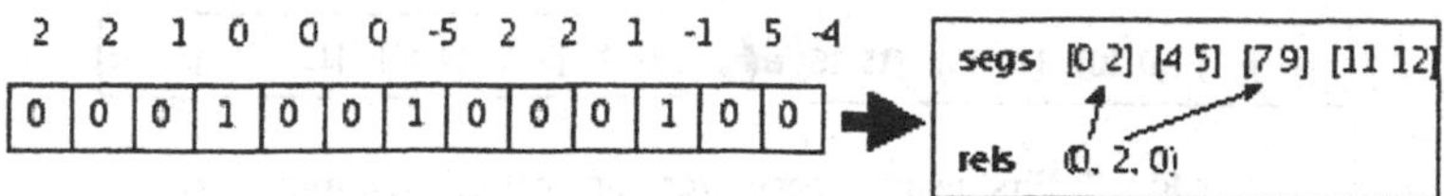

Fig. 2. An individual representing a segmentation of a musical excerpt and the relations created as a result of the evaluation process

tion is represented by a structure (seg_1_pos, seg_2_pos, *dif*), where seg_1_pos and seg_2_pos are the ordinal positions of the two segments (we consider that the first segment's position is 0) and *dif* is the difference between them. Although, from the genetic algorithm point of view, relations aren't part of individuals, it is obvious that this information is essential for the appreciation of the merit each individual has. Thus, whenever possible, in the rest of the paper individuals will be followed by the relations existing between their segments.

The goal of the system is to find an individual for which the sum of the length of the segments participating in relations is the maximum possible and for which de sum of the differences attached to all the created relations is the minimum possible. Stated another way, the goal is to find an individual that maximizes the following function

$$f(x) = a \times \sum_{i=1}^{m} d(rel_i) + b \times \sum_{i=i}^{m} area(s_i)$$

where x is the individual to be evaluated, a is a negative constant, b is a positive constant, m is the number of relations created as a consequence of the segment comparison process, $d(rel_i)$ is the difference attached to relation rel_i calculated with the Wagner-Ficher algorithm [7], n is the number of segments of x, and

$$area(s_i) = \begin{cases} r_i = l_i + 1, & \text{if segment } s_i \text{ makes part of at least one relation,} \\ 0, & \text{if not} \end{cases}$$

is the length of each segment participating in at least one relation (the length of each segment is summed only once, even if it participates in more than one relation). The Wagner-Ficher algorithm measures the distance between two strings by the number of operations of insertion, deletion and substitution needed to transform one into the other. The difference value between two segments, under which they are considered similar, was defined empirically as 5, if the smaller segment has length greater than 20, and as 1/4th of the smaller segment length, if not. We admit that more work is needed to choose the appropriate values.

Besides the difference between two segments, another aspect that can prevent the creation of a relation between them is their length. Actually, we allow the definition of a minimum value to the segments length — which we call *min_pat_len* — under which a segment cannot be considered similar to any other segment. This doesn't mean, however, that segments with a length less than *min_pat_len* cannot exist (the parameters that actually bound segments length are *min_seg_len* and *max_seg_len*). The existence of the *min_pat_len* parameter allows us to define that, for example, in the first segmentation of the piece being analysed, only the bigger patterns can be identified, although small *rests*, which are segments that don't participate in any relation, can coexist with these patterns in the segmentation.

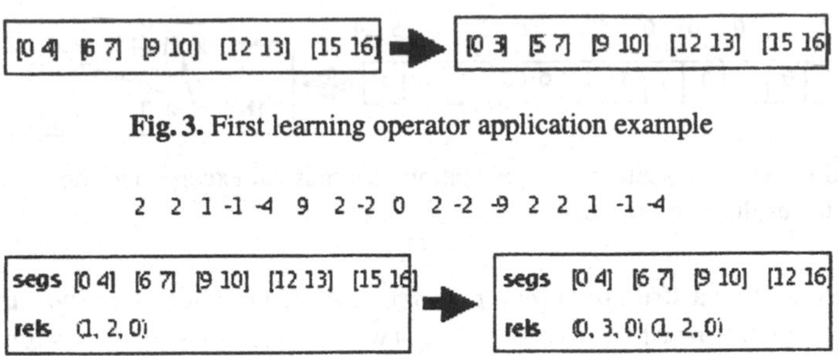

Fig. 3. First learning operator application example

Fig. 4. Second learning operator application example

A last word about evaluation: in the present version of the system, segments are only compared in their original form. This does not mean, however, that they cannot be also compared, for example, in their retrograde or inverted form. In fact, this will be one of the next steps we will take in improving the system.

2.4 Learning Operators

In the learning operators application step, four operators are applied to a small randomly chosen set of individuals. These operators are applied iteratively to each chosen individual until it is not possible to improve its fitness. At the end of each cycle the correction procedure described above is applied in order to correct possible invalid situations, but now at the phenotype level. This operation can, however, deteriorate the individual's fitness to a level worst than it had before the learning process. Consequently, the modifications done by the operators and the correction procedure only acquire a definitive character if at the end of the process an effective improvement is achieved. If this is the case, the individual's genotype is updated in order to reflect the modifications done at the phenotype level. We will now describe the four operators used in the learning process.

The first operator does the same thing to all limits between segments: first, it tries to shift the limit to the left the more units it can, which means that it stops once this implies a decrease in the individual's fitness. When this happens, it tries the same thing, but now to the right. Figure 3 illustrates the shifting in one unit to the left of the limit between the first and the second segments of an individual.

The second operator consists just in concatenating all the individual's consecutive *rests*. This operator, as the one we will describe next, was initially conceived with the single purpose of simplifying individuals' structure. However, its application can also cause an improvement in the individual's fitness since it is possible that it leads to the creation of new relations in which the new segments participate. Figure 4 illustrates this situation.

The third operator is better described through an example. Consider the musical excerpt of figure 5, which has an AXA structure. The individual allows the identification of all recurrent existing material, but it has a clearly more complicated structure than needed (ABCXABC). The purpose of this operator consists on the detection of these situations and on the simplification of individuals if this does not imply a fitness decrease.

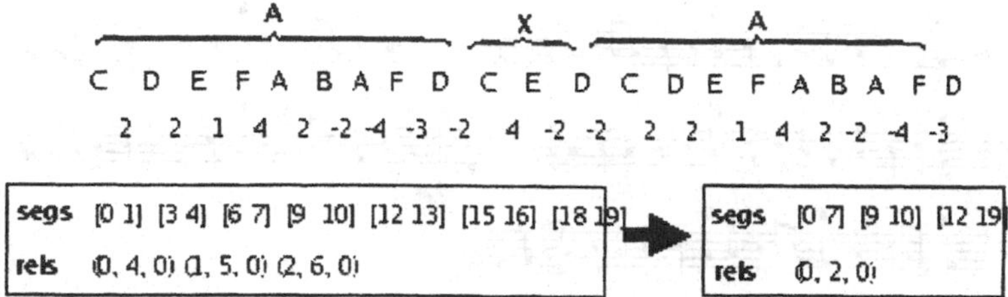

Fig. 5. Third learning operator application exampl

It can also be viewed as a bias for segmentations with longer patterns. Thus, although the fitness function does not state any preference for longer patterns this preference exists and is expressed through this operator.

The fourth operator is similar to the first one, but it operates simultaneously on the limits of segments that together participate in one relation. Consider, for example, that there exists a relation between segments $[l_x\ r_x]$ and $[l_y\ r_y]$. This operator starts by trying do shift the left limit of both segments to the left, resulting in the new segments $[l_x\text{-}1\ r_x]$ and $[l_y\text{-}1\ r_y]$ (these modifications imply that the limits of adjacent segments are also modified). This procedure is repeated while it does not decrease the individual's fitness. When this happens, the operator tries to shift left limits to the right the more positions it can. After left limits are treated, the same procedure is applied to the segments' right limits, but now, first to the right and then to the left.

2.5 Further Segmentations

It is consensual that almost all musical pieces have a hierarchical structure. This means that they are composed of bigger related segments that are also decomposable into smaller ones. In EMPE, further segmentations are done by first choosing some of the segments identified in the previous segmentation and then applying the same segmentation process to a "new musical piece" composed by these segments.

Although segment classification is not a central goal of our work, the problem of deciding which segments must be chosen to make part of the "new piece" to be analysed can be seen as the one of choosing the most representative segment - the prototype - of each segment class existing in the musical piece being analysed. The classification algorithm we used, which we will not describe in this paper due to lack of space, is rather rudimentary, being one of the main aspects we must improve in the future. One of its characteristics is that similarity relations between segments are considered as transitive, which is not true. This means that, if segment s_1 is similar to s_2, and s_2 to s_3, then all the three will belong to the same class, even if s_1 and s_3 are not similar. However, this was not a problem in all the experiments we have done, since in none of them a class was created to which two dissimilar segments belonged.

After the classification process, we must decide which of the segments of each class is its prototype. This decision is taken in the following way: if the class has just one segment, then that segment is its prototype; if it has two segments, choose the smaller of the two segments; if the class has more than two segments, its prototype is the segment to which the sum of the differences between it and the other segments of the class is smaller.

Fig. 6. "Maria muoter reinû mait"

The "new piece" to be analysed is created juxtaposing the chosen segments by the order they appear in the original piece. Between each two segments a special interval is inserted with the purpose of preventing, in the next segmentation, the creation of segments that are not sub-segments of the previously identified ones. This is achieved by placing a 1 in all individuals' genotype positions corresponding to special intervals and by prohibiting genetic and learning operators from modifying them.

3 Experimentation

In this section, we show the results obtained after some experiments done with the troubadour songs "Maria muoter reinû maît" (Fig. 6) and "Kalenda Maya"(Fig. 9), used by Ruwet in [8] to exemplify his analysis procedure. In these figures, we also show what segments this author identified. We have done 30 runs for each piece and, in each run, we used populations of 100 individuals evolving during 100 generations (the termination criterion is the number of generations since it is impossible to know, in advance, what is the fitness value corresponding to the best segmentation). The probability of applying crossover and mutation operators was defined, respectively, as 70% and 1%. In each generation, only one individual was submitted to the learning process. Values used for weights *a* and *b* of the fitness function were, respectively, -2 and 1. Parameters *max_seg_len* and *min_seg_len* were defined, respectively, as half the length of the piece and 2. Finally, parameter *min_pat_len* was defined as 10% of the length of the piece in the first segmentation and 2 in the second.

3.1 "Maria Muoter Reinû Maît"

In the first segmentation of this piece, all the best individuals generated in each of the 30 runs have a structure identical to the one depicted in figure 7, which corresponds to the first segmentation done by Ruwet. We can, thus, say that EMPE has no difficulty in discovering the higher-level structure of this piece.

The second segmentation was done using an interval sequence constructed from segments identified in the first segmentation. This sequence is composed by segments [36 39] and [71 88] (respectively, segments A' and B) by this order. Individuals equal to the one depicted in figure 8, which nearly correspond to the second segmentation

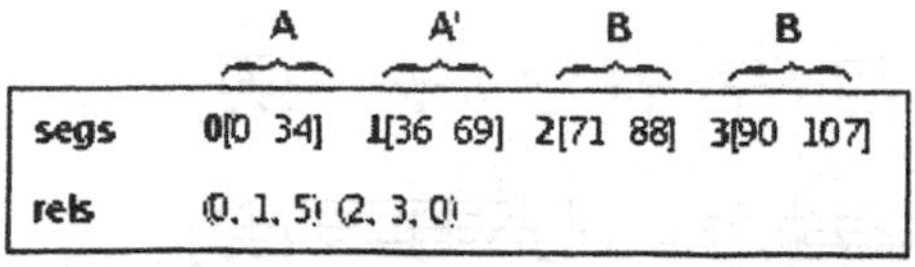

A A' B B

segs 0[0 34] 1[36 69] 2[71 88] 3[90 107]

rels (0, 1, 5) (2, 3, 0)

Fig. 7. Best individuals generated in the first segmentation of "Maria muoter reinû mait"

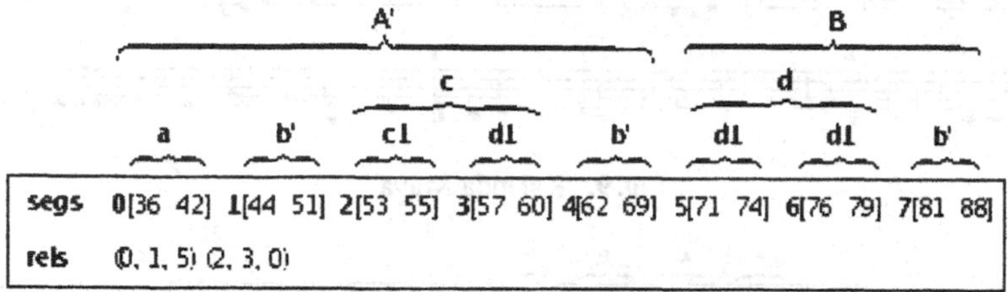

Fig. 8. Best individuals generated in the second segmentation of "Maria muoter reinû mait"

done by Ruwet, were generated in 21 of the 30 runs. The only differences are that segments c and d are both divided in two. However, it is only possible to identify the three occurrences of segment d1 by dividing them.

3.2 Kalenda Maya

In the first segmentation of this piece, 20 runs resulted in individuals with the same structure of the one in figure 10.

As is possible to confirm in this figure, these individuals' first four segments correspond to the first four segments identified by Ruwet. In order to explain the configuration of these individuals' last four segments we must first turn our attention into the two segments Ruwet identifies as D and D'. The difference between them, calculated with the Wagner-Ficher algorithm, is equal to 6, which is greater than $\frac{1}{4}$th of the length of segment D, the smaller of the two segments. This means that, even if an individual had two segments corresponding to segments D and D', no relation would be created between them during the evaluation process.

Given the impossibility of associating segment D and D' through a similarity relation, what does the genetic algorithm? Following the fitness function, it tries to find segments that lead to a bigger portion of covered area with the minimum possible differences between related segments. In order to achieve this, after the first four segments, it divides the rest of the piece in another four segments, two of which are equal (segments 4 and 6 in 10 These two segments, constituted by two occurrences of segment c plus the common beginning of segments D and D', are, after the two occurrences of segment A, the two bigger identical segments of the piece. This way of doing things is, in fact, consistent with Ruwet's analysis procedure: "first try to identify the bigger identical segments existing in the piece".

In the second segmentation of this piece, the interval sequence that was analysed was composed by segments [0 20], [44 52], [64 73], [75 77] and [71 88] by this order. Figure 11 depicts the structure of the 18 best individuals generated in the 30 runs done. These individuals, besides the two occurrences of segment c, allow the identification of

Fig. 9. "Kalenda Maya"

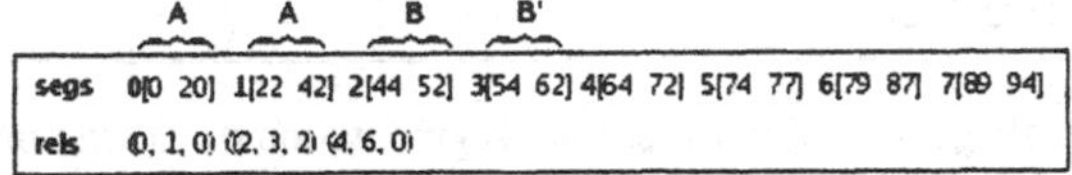

Fig. 10. Best individuals generated in the first segmen

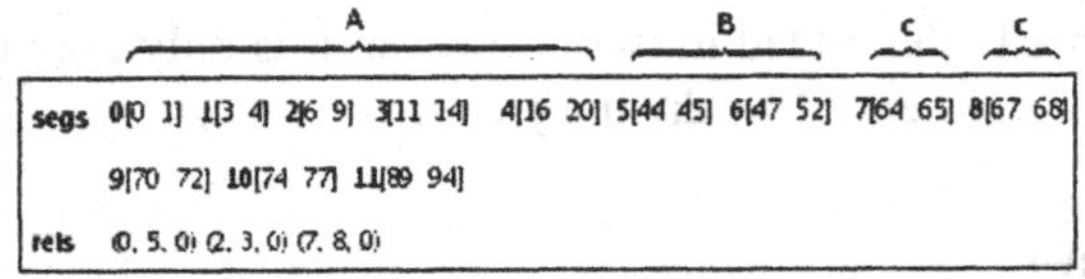

Fig. 11. Best individuals generated in the second segmentation of "Kalenda Maya"

the similarity between the beginning of segments A and B/B' (segments 0 and 5 in the figure), and the similarity between two sub-segments of A (segments 2 and 3).

4 Conclusions and Future Work

In this paper we have presented a genetic algorithms based approach to musical pattern extraction. In the first phase of this research work we have been more engaged with the question of how this type of algorithms can be used to effectively solve the problem without worrying about complexity issues. The experiments we have done allow us to think that, in fact, this approach is a viable one (besides " Maria muoter reinû maît" and "Kalenda Maya", we have done experiments with some others pieces as, for example, Debussy's "Syrinx", with also good results).

Our future work will comprise two types of tasks: some in which the goal will be to increase the quality of the generated segmentations, as well as the probability of generating them, and some in which the goal will be to study and, if possible, reduce the algorithm's complexity (each run for the first segmentation of the two pieces used in this paper took approximately 0.4 seconds in a AMD Athlon XP 1.54 GHz). Some ideas for the first set of tasks are: the utilization of more music oriented distance measures such as the ones described in [2] or [9]; the comparison of segments in their retrograde and inverse form; the utilization of a better classification algorithm. In what concerns to complexity, it can be reduced if we avoid the comparison of two segments more than once. This may happen because it is almost certain that, during the search process, two or more individuals will have some common segments. We think that a data structure similar to the one used in [2] may help avoiding this repetition. The challenge with this method will be to maintain space complexity to reasonable levels.

References

1. Rolland, P.Y, Ganascia J.G.: Musical Pattern Extraction and Similarity Assessment. In E. M. Miranda (Ed.), *Readings in Music and Artificial Intelligence*, Harwood Academic (2000) 115 -144
2. Rolland, P.Y.: Discovering Patterns in Musical Sequences. *In Journal of New Music Research, 28, n° 4* (1999) 334-350
3. Cambouropoulos, E.:*Towards a General Computational Theory of Musical Structure.* PhD Thesis, University of Edinburgh (1998)
4. Meredith, D., Wiggins, G. A., Lemström, K.: Pattern Induction and Matching in Polyphonic Music and Other Multi-dimensional Datasets. *Proceedings of the 5th World Multi-Conference on Systemics, Cybernetics and Informatics (SCI2001),* Volume X (2001) 61-66
5. Mitchell, M.: *An Introduction to Genetic Algorithms,* MIT Press (1996)
6. Pereira, F.:*Estudo das Interações entre Evolução e Aprendizagem em Ambientes de Computação Evolucionária.* PhD Thesis, Universidade de Coimbra (2002)
7. Stephen, G.:*String Search.* Technical Report, School of Electronic Engineering Science, University College of North Wales (1992)
8. Ruwet, N.: *Langage, Musique, Poésie.*Editions du Seuil, Paris (1972)
9. Perttu, S.:*Combinatorial Pattern Matching in Musical Sequences.* Master's Thesis, University of Helsinki, Series of Publications, C-2000-38(2000)

Automatic Extraction of Approximate Repetitions in Polyphonic Midi Files Based on Perceptive Criteria

Benoit Meudic[1] and Emmanuel St-James[2]

[1] Musical representation team - IRCAM
1, place Igor-Stravinsky, 75004 Paris, France
meudic@ircam.fr
[2] LIP6/SRC, Université Pierre et Marie Curie
4, place Jussieu, 75255 Paris, France
Emmanuel.Saint-James@lip6.fr

Abstract. In the context of musical analysis, we propose an algorithm that automatically induces patterns from polyphonies. We define patterns as "perceptible repetitions in a musical piece". The algorithm that measures the repetitions relies on some general perceptive notions: it is non-linear, non-symetric and non-transitive. The model can analyse any music of any genre that contains a beat. The analysis is performed into three stages. First, we quantize a MIDI sequence and we segment the music in "beat segments". Then, we compute a similarity matrix from the segmented sequence. The measure of similarity relies on features such as rhythm, contour and pitch intervals. Last, a bottom-up approach is proposed for extracting patterns from the similarity matrix. The algorithm was tested on several pieces of music, and some examples will be presented in this paper.

1 Introduction

Automatic music analysis is an increasingly active research area. Among the main tackled subjects, the search for *musical patterns* is at a central place. Indeed, most of the musical pieces are structured in various components ("phrases", "motives", "themes", "generative cells"...) that can naturally be associated with the notion of pattern. Considering the only notion of "pattern" simplifies (or postpones) the issue of making the distinction between the different natures of the components (theme, motive etc...) of a musical piece. However, answering to the question "what is a pattern?" is still rather difficult. Often, the notion of pattern can be linked with the notion of repetition: patterns emerge from repetition. But a pattern could also be defined by its salient boundaries, and then patterns would emerge from discontinuities in the music. Last, patterns can also be characterized as independent and coherent entities. Providing a definition is all the more difficult as we place ourselves in a musical context. Most of the time, patterns are linked with perceptive notions, which raises one question: can we consider as "patterns" the structural (one would say mathematical) regularities of a musical sequence, even if theses regularities are not perceived? Inversely, do perceived repetitions only correspond to exact repetitions? We think that the notion of similarity between two sequences plays an important role in the perception of patterns and it should be part of a pattern extraction system.

U.K. Wiil (Ed.): CMMR 2003, LNCS 2771, pp. 124–142, 2004.

In this article, we have chosen to focus on perceptible musical structures. Moreover, we assume that musical structures can be induced from the extraction of repeated sequences (that we call patterns) and thus we address the issue of extracting "perceptible repetitions" from a musical piece.

2 Background

The literature is quite poor in algorithms that automatically extract patterns (perceptible repetitions) from polyphonic music.

An interesting method, starting from the audio, is proposed by Peeters [1]. It considers the signal as a succession of "states" (at various scales) corresponding to the structure (at various scales) of a piece of music. A similarity matrix is computed from feature vectors. The similarity matrix reveals large and fast variations in the signal that are analysed as boundaries of potential patterns. The patterns are then grouped according to their similarity. The method is relevant for pieces that contain salient transitions (timbre changes, important energy variations etc...), but could reveal less relevant for the detection of phrases in piano music.

Another method in Lartillot [2] starts from MIDI files and induces patterns by analysing the musical sequence in a chronological order. All the possible combinations of successive events contained in a temporal window of limited size are potential candidates for being a pattern. The notions of temporal context and expectation are modelized. However, if very promising, this method cannot analyse long sequences with too many events because it would require a too high computation cost. Moreover, polyphonic context (see 4.1 for definition) is not considered.

Rolland [3] proposes a model that uses dynamic programming. All pairs of possible patterns of a given sequence are compared. Pairs whose similarity value overlaps a given threshold are memorized into a similarity graph. An interesting optimisation of the computation time is proposed. However, the similarity measure that is used is linear which might not be cognitively relevant for us as we consider a pattern as a whole entity (see equation 1 in 5.2). Moreover, the algorithm cannot extract patterns from unvoiced polyphonic music, such as piano music.

Meredith et all [4] propose a multidimensionnal representation for polyphonic music. The algorithm searches for different matchings of the geometrical figures of feature vectors by translating them in a multidimensionnal space. This approach is interesting because it offers a graphical answer to a musical issue. However, only exact repetitions can currently be extracted. The way to solve this issue seems to find relevant representations that could be identical for musical patterns that are perceptively similar but physically different. Otherwise, the repetition detection algorithm should be modified.

Last, Cambouropoulos [5] proposes to segment a MIDI file and then to categorize the segments in clusters. The segmentation algorithm uses a boundary detection function that computes the similarity between all the possible patterns of the sequence. Then, the segments are clustered according to a threshold that depends on the different distances between the segments. Once the clusters are computed, the distances between segments are computed again in order to optimize the current clustering, and the clustering function is called again until the clusters are found optimal. This method is

Fig. 1. Beginning of the Sonata for piano No9 EM from Beethoven. The first note and the three chords form a pattern. Transforming the three chords into three independent voices would not be perceptively relevant.

interesting because it analyses whole sequences of polyphonic MIDI music, and the notion of context is originally used in one of the two steps of the analysing (the clustering). However, the segmentation step appears hazardous, and would require a high computational cost.

3 Aims

The algorithm we present in this article is a new model for extracting patterns from polyphonic MIDI sequences. Our aim is that the patterns we extract can be considered as components of the musical structure. In order to build a system the more general as possible, we do not consider any knowledge based on tonality neither we consider particular styles. Several perceptive notions are taken into account in the algorithm. However, in a first step, we have chosen to challenge the limits of a system that does not modelize temporal context or expectation. Indeed, the integration in the system of all the cognitive mechanisms that play a role in our perception of music is far too complex. Thus, we have to draw the limits of the model. The consideration of temporal context and expectation should be possible in a second step, but we first prefer to explore the limited system, and we will rely on the results to show that our method offers promising applications. Besides, we will not try to extract all but a set of significant patterns from polyphonic music. If our aim is that the extracted patterns are the most relevant of all the sequence, we consider as more important the relevance of the pattern itself.

4 Introduction to the Pattern Extraction Model

4.1 General Considerations on Patterns

We have defined patterns as "perceptible repetitions in a musical sequence". This notion has to be refined before we describe in details the model.

An issue arises when refining the term 'repetition' of the definition. How to define the similarity between two sequences? An attempt to answer to this question is proposed in 5.2.

Another issue stems from the polyphonic context: is a pattern a single melodic line inside a polyphony, or is it itself a polyphonic component of a polyphony? For instance, let's consider the following extract of the 9nth Beethoven's sonata (Fig. 1). The sequence that is composed by the first four events can be called a pattern as it is repeated

Fig. 2. A pattern extracted from the Sacral dance of the Rite of the spring from Stravinsky. A monodic voice could hardly be isolated from the polyphonic pattern.

several times. This sequence is polyphonic (one note followed by three chords). Should we consider that the sequence of chords is composed by three different monophonic patterns that are superposed, or is it only one polyphonic pattern? Would it be relevant to dissociate the chords in as many components as there are notes inside, or should we consider them as entities? This also asks the question of the independence of a melodic line from its polyphonic context. Can we always isolate one or several monophonic lines, or do the superposition of several notes form a unique entity that cannot be decomposed? In several musical examples, there would be no reason to extract a single monodic line (see also Fig. 2).

We also claim that considering a single melodic voice without its polyphonic context is often nonsense, as the polyphonic context often plays an important role in the perception of the structure of the piece. For instance, if we consider the beginning of the Sonata AM d664 op121, 1rst Part from Schubert (Fig. 3), one can extract the melody from the voice at the top. However, the segmentation of the melody in different patterns (that correspond in musical analysis to the usual "antecedent-following" repeated one time with variation) is not trivial without information on the polyphonic context. Similarly, if we consider the beginning of the Intermezzo op117 no1 from Brahms (Fig. 4), we could extract the melody by hand with difficulties (it is not at the top of the voices, and it is repeated but with big variations), but it would not characterize the excerpt. Indeed, one could imagine the same simple melody in very different polyphonic contexts, divided in several different voices.

However, and this raises one of the main issues, the only melodic line should also be taken into account, for instance when it follows the model 'melody + accompaniment'. Indeed, in this case the melodic line could be separated from the accompaniment and considered as a monophonic pattern that would be repeated in the following music sequence, but with different accompaniment, or with melodic variations. This is mainly a perceptive issue that is difficult to solve with only one rule.

Sometimes, the solution is trivial. For instance, melodic lines are often difficult to follow in Bach's Fugues, but fortunately, they appear at the beginning of the fugue one after the other, which let us time to memorize them. In this case, the pattern first appears in a monophonic context (the Fugue's subject) and then is repeated and varied in a polyphonic context. The issue is not to extract a pattern (monophonic or polyphonic) from polyphony but to recognize a known (or memorized) monophonic pattern inside polyphony. However, in some pieces such as canons, only the beginning of the pattern appears in a monophonic context. The repetition of the pattern covers a part of it. Thus,

Fig. 3. Beginning of the Sonata AM d664 op121, 1rst Part from Schubert. The four patterns correspond to the structure antecedent1-following1-antecedent2-following2. This structure could not be extracted from the consideration of the only melodic line.

in a polyphonic context, music cannot always be segmented in individual successive segments (that is often proposed in several musical theories) and possible coverings between the structural components must be taken into account.

Whatever the situation would be, we believe that the polyphonic context plays an important role. Even if a single melodic line could be extracted from the polyphony, the polyphony should be associated to the melody in most of the cases. One could say that when remembering a polyphonic excerpt, we often sing a single melodic line, but this is due to our physical impossibility to sing a polyphonic sequence, and often when singing the melodic line, we hear (but don't sing) the polyphonic context (at least the harmony) that was associated with it. It means that we have memorized it, and that we take it into account in our comparisons with other sequences. In this article, we propose to define a pattern as a polyphonic component of a polyphonic sequence.

4.2 General Architecture of the Model

Our model analyses a MIDI file that contains the onset (in milliseconds), the pitch (in midicents) and the duration features. The dynamic and the channel features are not considered.

Fig. 4. Beginning of the Intermezzo op117 no1 EM from Brahms. The sequence can be divided in four similar patterns (1-6, 7-12, 13-18 and 19-24) Polyphonic context is part of the structure of the sequence. The only melody line with another polyphonic context would not be stated as similar to this one.

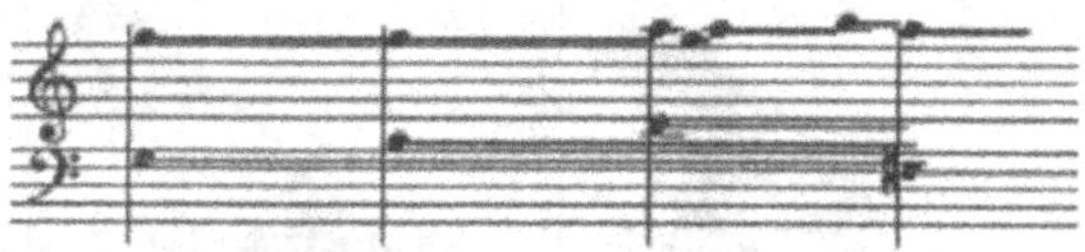

Fig. 5. A sequence of beat-segments (b.s) extracted from the "Variations Goldberg" from Bach. The vertical lines delimit the beat segments. Horizontal lines are the durations of each event.

The model is composed of two main algorithms:

- the first algorithm computes several similarity matrices from a quantized MIDI file 5.
- the second algorithm extracts patterns from the similarity matrices 5.1.

In the first algorithm, we first quantize and segment the MIDI sequence with an algorithm proposed in Meudic [6]. The boundaries of the segments correspond to the downbeats of the music. The initial sequence of MIDI events is thus considered as a sequence of beat-segments. Each beat-segment (b.s) is itself a sequence of MIDI events (see Fig. 5). Then, given a length L (in number of beats), each sequence of b.s of length L is compared with all the other sequences of same length. For each comparison, three similarity values are computed (corresponding to three different features: pitch intervals, pitch contour and rhythm) and associated with the events of the two sequences that have been found similar (we call them "templates"). All the similarity measures are then stored in similarity matrices. The measures of similarity we use are presented in 5.2.

In the second algorithm (5.1), patterns are extracted from the similarity matrices. First, the matrix cells (a matrix cell is a couple of two sequences of b.s of same length) are filtered and clustered according to their similarity value. Then, patterns are extracted in a bottom-up approach that starts from the different clusters of cells and then groups them in new (and often smaller) clusters with longer cells.

5 The Similarity Matrices

5.1 The Computation of a Matrix

A similarity matrix (Fig. 6) stores the similarity values between the pairs of b.s (beat.segment) sequences of same length L (in number of beats). The units of the vertical and horizontal axis are expressed in number of beat-segments. We consider the

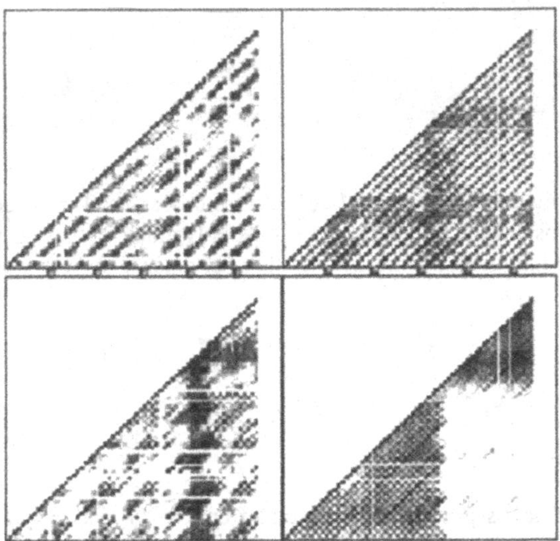

Fig. 6. Four different similarity matrices from first 30 seconds of (from left to right and top to bottom): Sonata AM d664 op121 2nd Part from Schubert, 1rst Gymnopedie from Satie, 1rst and 3rd part of the Sacral dance of the Rite of the Spring from Stravinsky. White areas correspond to dissimilar cells. The unit for both vertical and horizontal axis is expressed in number of beat-segments (one cell per beat-segment).

matrix as symmetric (see discussion in 5.2). Each cell refers to a pair of b.s sequences. The vertical and horizontal positions of each cell are the two beginning positions (in number of beats) of the two corresponding b.s sequences.

Each similarity measure provides a real value between 0 and 1 that states how similar are the compared sequences (1 is for identical).

In the model, sequences of b.s are of same length (length is expressed in number of b.s) so that each position of b.s in a given sequence can be matched with the same position of b.s in the other sequence.

The choice of the length L can appear somewhat arbitrary. The issue is to find the "right level" between the maximum length of patterns that cannot be divided in several parts and the minimum length of patterns that can be divided. Indeed, we think that some patterns must be considered as whole entities (see sub-section 5.2) and not as the concatenation of smaller patterns. Two such patterns (sequences of beat-segments) can be similar while their components (the beat-segments), if considered individually, are different. We think that the maximal length of such patterns could be linked to the limits of our short-term memory: when listening to a musical sequence, we compare what we are hearing with what we have already listened. If what we are hearing is a variation of something already listened, we will match the two sequences. We match the sequences because they share common aspects. The question is: what is the minimal number of common aspects, and the maximal distance between the aspects, that is required to initiate a match between the sequences. In other words, what is the maximal size of a pattern (for us, the size is in number of beat-segments) that must be considered for matching it with another pattern?

Fig. 7. Seq1 and Seq3 have similar consecutive intervals but are globally different. Seq2 and Seq3 have also similar consecutive intervals but are globally more similar. Intervals between not consecutive events have to be considered to measure this difference.

In our tests, we have decided to define a "blurred" length of pattern L: we compute the similarity between pairs of sequences of length L, L-1, L-2 and L-3, and we choose the best (the higher) similarity value (a decreasing coefficient is applied to the similarity value when the length of the sequence decreases, in order to support longer sequences).

5.2 The Similarity Measures

In this part, we define several similarity measures between two b.s sequences of given length.

We compute three different similarity values by considering three different sets of features: pitches (chords, pitch intervals etc...), pitch contours (contour at the top and at the bottom of the polyphony) and rhythm. The similarity values are computed in respect with some cognitive aspects (see sub-section 5.2). Each time a similarity value is computed between two sequences seq1 and seq2, it is associated with two "templates", that is to say the events of seq1 similar to seq2 (template1) and the events of seq2 similar to seq1 (template2). These templates will be used to refine the kind of similarity relation that exists between the two sequences (see sub-section 6.1).

Similarity in a Polyphonic Context. We assume that the notion of similarity between two polyphonic musical sequences makes sense. No information is available on the different voices of each sequence. Computing the intersection between the two sequences would appear as an intuitive way to measure what is common between the two sequences. However, the intersection could be empty while the two sequences would be perceived as similar.

Thus, we state that a sequence x is similar to a sequence x^1 if x is approximatively included in x^1. For instance, when listening to music, we try to associate one sequence already heard with the current sequence we are hearing. We do not intersect the two sequences, but we evaluate the similarity between one sequence and a reference one. Thus, in our model, we understand similarity between two sequences x and x^1 as the distance from x to a certain sequence $sub(x^1)$ included in x^1.

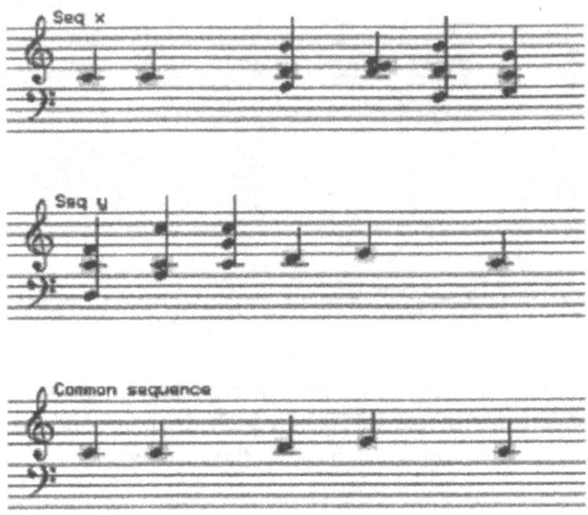

Fig. 8.

Cognitive Aspects of the Similarity Measure. First, a musical sequence of b.s is considered as a whole entity (it may contain an abstract cognitive structure) and not solely as the concatenation of smaller entities. Thus, the similarity measure is not linear (see equation 1). Indeed, we think that it is necessary to consider the relations between non-adjacent events. Theses relations play a role in the cognitive process for recognizing the similarity between two sequences.

For instance, two sequences that are locally similar but that have different global pitch range may not be very similar (see Fig. 7).

$$S(x,x^1)+S(y,y^1)\neq S(xy,x^1y^1).$$

($S(x, x^1)$ designs the similarity value between sequences x and x^1, and xy designs the concatenation of sequence x and y)

Another cognitive aspect is that our similarity measure is not symmetric in a polyphonic context (equaltion 2).

$$S(x,x^1)\neq S(x^1,x).$$

In our model, a sequence x is similar to a sequence y if x is approximatively included in y. However, this do not mean that y is included in x and thus we do not know if y is similar to x. Moreover, if the first half of x is included in y, and the second half of y is included in x, the similarity cannot be determined. To compute the similarity, we must determine either if x is entirely included in y or if y is entirely included in x (see Fig. 8).

Last, according to Tversky [7], the similarity measure is not transitive, that is to say the triangular inequality is not true (equaltion 3).

$$S(x,y)+S(y,z)\leq or \geq S(x,z).$$

For instance, z can be a variation of y that is a variation of x. But z can be very different from x and thus not judged as a variation of x.

Fig. 9. Two similar patterns from Sonata DM d664 op121 2nd Part from Schubert. A reference pattern (the above one) is compared to another pattern (the similarity measure for pitches is 0.8 on 1). Durations are not represented. The high similarity can partially be explained by the fact that: some chords are repeated (C1 is similar to C2, and C1' is similar to C2'), C3 is similar to C3', the melodic line m1 is similar to m1', and the melodic line m2 is similar to m2'.

The three different cognitive aspects have been considered in our model of similarity. We will now provide more details on the different similarity measures.

Similarity Measure for Pitches. In this section, we consider the chords and the pitch intervals features. A similarity value is computed from two b.s (beat.segments) sequences seq1 and seq2 of same length.

The only events falling on the downbeats are considered. This may be arguable, but two reasons have conducted this choice:

- considering all the polyphonic events would require too much running time
- the downbeats are often perceived as salient temporal position. Two sequences whose pitches coincide on the downbeat but differ elsewhere are often recognised as very similar (this has been confirmed in our experiments).

Usually, a downbeat event (dwb.event) is a chord, but it can also be a note or a rest. We focus on two musical aspects: the vertical and the horizontal one.

1. The vertical aspect concerns the repetition of chords. A chord is represented by the intervals between the lower pitch and the other pitches of the chord. The intervals are computed modulo 12.
2. The horizontal aspect concerns the similarity between melodies. A melody is any sequence of consecutive pitches.

5.2.3.1 The Repetition of Chords. In order to consider all the possible relations between non-adjacent chords, the global similarity results from the mean of the similarity values between all the pairs of chords in seq1 and their corresponding pairs in seq2 (see Fig. 9).

The similarity function between a pair P1(C.1, C.1') and a pair P2(C.2, C.2') is provided in the following listing.

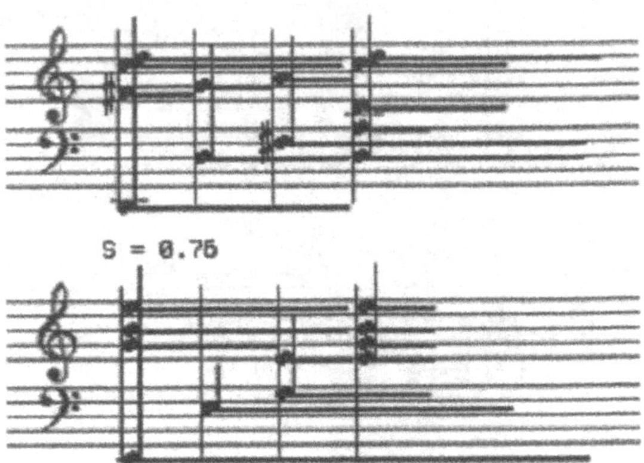

Fig. 10. Two similar patterns from intermezzo op119 n°3 from Brahms. The contours at the top are similar (contour D-D and contour E-E). Pitches in the middle of the two sequences are not in the contours because the durations (horizontal lines) of the first pitches D and E of each contour cover them.

Listing in Lisp of the similarity function between two pairs of chords C.1, C.1' and C.2, C.2':

```
(defun Sim(C.1 C.1' C.2 C.2')
     (max (mean (/ (length (inter C.1 C.1'))
                      (min (length C.1) (length C.1'))))
                 (/ (length (inter C.2 C.2'))
          (min (length C.2) (length C.2')))))
     (min (/ (length (inter C.1 C.2))
                      (min (length C.1) (length C.2))))
                 (/ (length (inter C.1' C.2'))
                      (min (length C.2) (length C.2')))))))
```

First, the function computes two values (see fig. 9): the percentage of the common notes between the chords in the pair (C1, C1') and the chords in the pair (C2, C2'). If C1 and C1' have common notes and C2 and C2' have not, the pairs P1 and P2 are still perceived as similar. Thus we consider the mean of the two values.

Then, the function computes another two values: the percentage of the common notes between the chords in the pair (C1, C2) and the chords in the pair (C1', C2'). If C1 and C2 have common notes and C1' and C2' have not, the two pairs can still be very different. Thus, we consider the min of the two values instead of the mean.

5.2.3.2 The Similarity between Melodies. Considering two sequences of beat-segments seq1 and seq2, a function computes the best fitting between seq1 and different transpositions of seq2. The aim is to find melodic lines in seq2 that are included in seq1 (see Fig. 9). To simplify the description of the model, we consider that each down-beat of each beat-segment correspond to one chord. Thus, we compare two sequences of chords seq1 and seq2. Seq1 and seq2 have same number of chords L.

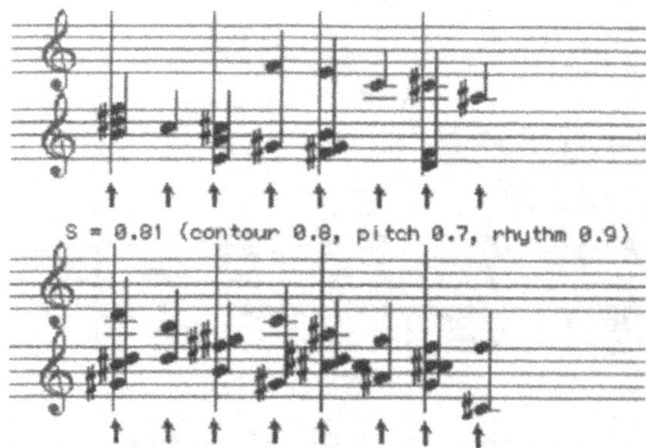

Fig. 11. Two similar patterns from Pierrot Lunaire from Schoenberg. Rhythm is very similar. Corresponding onsets are marked with an arrow.

The model contains L-1 iterations. Each iteration i compares sequences seq1 and seq2 whose first i-1 chords have been deleted. The same comparison function is processed for each iteration. This function outputs a similarity value between 0 and 1. The global similarity value between the two sequences seq1 and seq2 is the mean of L-1 similarity values that are computed for each iteration.

We will now describe the comparison function for the first iteration. Considering one pitch of the first chord of seq1, pitch(i)(chord1(seq1)), we try as many transpositions of seq2 as there are pitches on the first chord of seq2. A transposition consists in transposing the pitches of seq2 so that one pitch of (chord1(seq2)) is equal to the pitch pitch(i)(chord1(seq1)). The length (in number of chords) of the longest intersection of seq1 with the different transpositions of seq2 is memorized. This length is divided by the total number of chords L (remind that seq1 and seq2 have same length L). The iteration of the process to the N pitches of the first chord of seq1 provides N values between 0 and 1. The final result of the comparison function is the mean of the N values. Then, the comparison function is applied to the second chord of seq1 and seq2.

Similarity Measure for Contours. Our model compares the upper and lower contours of two b.s sequences seq1 and seq2 of same length (see Fig. 10).

As above, the only events falling on a downbeat (dwb.events) are considered. An up (down) contour is the sequence of the upper (lower) pitches of the dwb.events. If the duration of a pitch p1 covers the duration of the following pitch p2 and if p1 is higher (lower) than p2, then p2 will not be part of the up (down) contour.

Each contour of each sequence is compared with the two other contours (up and down) of the other sequence. The similarity value between two contours is the mean of the similarity values between all possible pairs of intervals between corresponding pitches of the two contours. The numerical values we use for comparing two intervals int1 and int2 are the following:

Fig. 12. Three patterns similar to the above pattern in Sonata AM d664 op121, 1rst Part from Schubert. The similar patterns are sorted according to their decreasing global similarity value S. The square symbols only determined the similarity for rhythm. The oval symbols determined similarity for both pitches and contour.

- 0.7 for down contours if int1 = int2
- 1 for other contours if int1 = int2
- 1 for up contour if sign(int1) = sign(int2) and abs(int1 - int2) ¡4
- 0.7 for other contours if sign(int1) = sign(int2) and abs(int1 - int2) ¡4
- 0.4 if sign(int1)*sign(int2) = 0 and if abs(int1 - int2) ¡ 3
- 0 if sign(int1)*sign(int2) = 0 and if abs(int1 - int2) ¿= 3
- 0.1 if abs(int1 - int2) ¿ 9
- 0.3 if abs(int1 - int2) ¿ 3
- 0 else

Similarity Measure for Rhythm. Our model compares the rhythmic structure of two sequences of b.s seq1 and seq2 of same length (see Fig. 11). In a first step, seq1 and seq2 are normalized so that the total duration of the b.s will be the same for seq1 and seq2. Then, for each b.s, onsets (temporal positions) in seq1 are associated to the corresponding onsets in seq2. Two onsets of two b.s form a pair if they share similar temporal positions in the b.s. If an onset of one sequence does not form a pair with an onset of the other sequence, then it is deleted.

The similarity between two sequences of b.s is the mean of the similarity between each corresponding b.s (as seq1 and seq2 have same length, each b.s of seq1 correspond

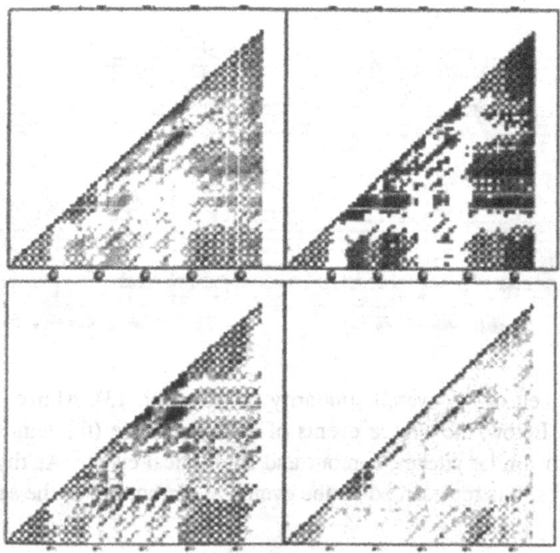

Fig. 13. Four similarity matrices from Sonata for piano N°9 EM from Beethoven (from left to right and top to bottom): overall similarity measure, similarity for rhythm, similarity for contour and similarity for pitches. The three measures are very different (see Fig. 15 for comparison). The matrix corresponding to rhythm is less informative than the other matrices because dark diagonals do not clearly emerge.

to one b.s of seq2). The similarity between two corresponding b.s is the mean of the the addition of the similarities between each pair of corresponding onsets divided by the total number of onsets of the reference b.s. The similarity between corresponding onsets is proportional the length of the intersection of the corresponding durations (an approximation value of the durations is considered for the intersection).

Overall Similarity Measure. Each of the three above measures (pitches, contour and rhythm) computes a similarity value between each pair of sequences of length L contained in the musical sequence we analyse. The results can be represented in three different similarity-matrices that can be analysed separately. Sometimes, it appears that rhythm, pitches and contour play a different role in the similarity measure, and sometimes the similarity matrices for the different measures are very similar (see Fig. 13 and Fig. 15).

The three measures can also be linearly combined into a global similarity measure (see Fig. 12 and Fig. 14 for similarity between sequences and Fig. 13 and Fig. 15 for similarity matrices). In this case, different weights can be applied to the different measures. An algorithm could be used to determine the best weightings by comparing the output of the similarity measures to a set of expected similarity-matrix. In our experiments, we have chosen to give the same weight to the different measures.

Due to the non-symmetrical relation (see equation 2), the similarity value between two non-ordered sequences seq1 and seq2 is composed of two different values: S(seq1, seq2) and S(seq2, seq1). In our model, we have chosen not to consider the two different

Fig. 14. Detail of one cell of the overall similarity matrix (Fig. 13). Above, the two compared sequences of the cell. Below, the similar events of each sequence (the templates). Numbers 1, 2 and 10 correspond to similar pitches, contour and rhythmical events. As rhythmical similarity do not considers pitches, it is represented by the events at the bottom of the sequence with static pitch C.

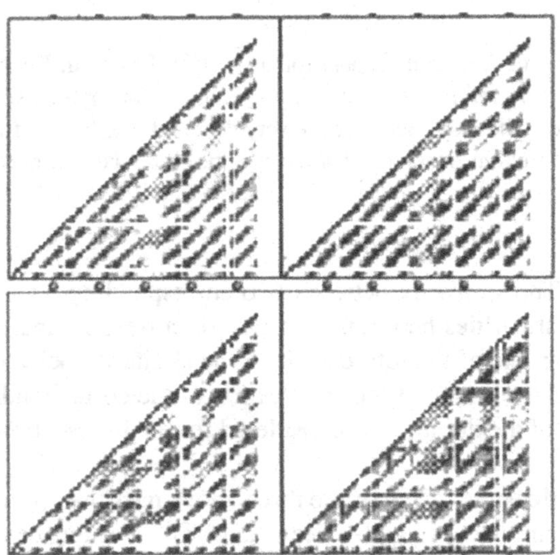

Fig. 15. Four similarity matrices from Sonata DM d664 op121 2nd Part from Schubert (from left to right and top to bottom): overall similarity measure, similarity for rhythm, similarity for contour and similarity for pitches. The matrices of the three different measures are very similar (see Fig. 13 for comparison).

values and thus we only consider the greater value. Doing that, our similarity matrix is symmetric and we represent only half of it.

6 Pattern Extraction from the Similarity Matrix

Several different types of information can be extracted from the matrices: general evolution for rhythm, pitch intervals or pitch contours, local repetitions of cells, areas of structural changes, etc...

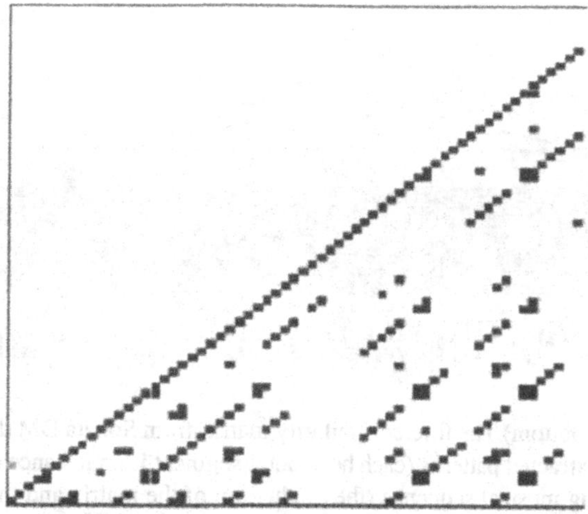

Fig. 16. A filtered similarity matrix from Sonata DM d664 op121 2nd Part from Schubert (the initial matrix is in Fig. 15). The most similar patterns appear (diagonal lines).

All of theses informations could be linked with the notion of pattern. However, in this part, we will only focus on the extraction of "the most important" patterns.

Defining such patterns is quite difficult, as there are no objective criteria to characterize them. One would agree to say that the "most important" patterns are the ones that are perceived as the "most salient" in the musical sequence: thus the attribute "most important" is related to perceptive criteria.

We think that the musical temporal context would play an important role in the definition of those criteria but once again, we will try not to consider it. Two other criteria are often found in the literature: the length of the pattern, and the number of its repetitions. However, it is quite difficult to combine the two criteria: the "most important" patterns often appear as a compromise between the two.

For instance, the same accompaniment can be found all along a musical piece. This pattern is very often repeated and thus very important in regards with the second criterion. However, it rarely appears to us as the "most important" pattern. Inversely, musical pieces that are exactly repeated two times can be seen as two very long patterns but they are not the most relevant for us because the repetition is trivial.

We will now propose a method for extracting patterns that could be "the most important ones". First, we propose to filter and cluster the cells of the matrix. Then, we propose to concatenate the cells in long patterns.

6.1 The Filtering/Clustering of the Cells of the Similarity Matrices

As we are looking for repetitions, we focus on the patterns that are very similar. For that, we choose a threshold that selects the only cells with the higher similarity values (see Fig. 16).

Then, we cluster the cells: each horizontal line of the similarity matrix represents all the similarity values between a reference sequence s(ref) and all the sequences of the

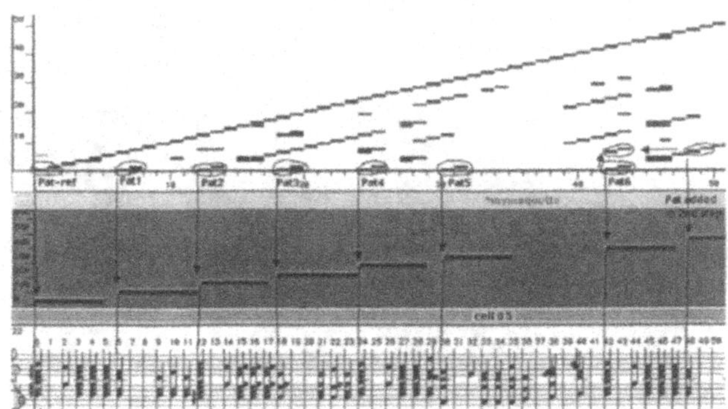

Fig. 17. (from top to bottom) The filtered similarity matrix from Sonata DM d664 op121 2nd Part from Schubert, the extracted patterns (each horizontal segment is an instance of the same pattern) and the corresponding musical sequence (the graduation of the matrix and the musical sequence is in "number of beat-segments").

line s(i). The high similarity values reveal high similarity with the reference sequence, but they do not reveal the kind of similarity, which has to be evaluated. For that, we compare the templates of the sequences. The template of seq1 (respectively seq2) contains the events of seq1 (resp seq2) similar to seq2 (resp seq1) (see Fig. 14). Sequences that have very similar templates are considered as sequences that have same kind of similarity.

6.2 The Extraction of Patterns

We define a bottom-up approach for pattern extraction that starts from the clustered cells and builds new patterns by concatenation of the consecutive cells of a same diagonal line.

The algorithm proceeds along the matrix from the lowest horizontal line to the upper one. For instance, in Fig. 17, the lowest horizontal line contains 8 cells. Only 6 of theses cells are clustered together. If we consider the second horizontal line, 6 cells can be concatenated to the 6 clustered cells of the first line. Thus, we have extracted 6 patterns of length "2 cells". As there are no cells in the third line, we stop the process.

In a second step, we associate to each pattern the corresponding horizontal lines of the matrix (see 6.1). Theses lines contain clustered cells that could be concatenated and that could form new patterns (see Fig. 17). Theses patterns could then be added to the six other ones, depending on their kind of similarity with them. In Fig. 17, we highlight the 6 extracted patterns and a pattern that has been added in a second step (the last one on the right of the figure).

We have tested several different musical pieces, and patterns that are part of the musical structure were often extracted. All of the musical examples that have been presented along the paper have been analysed, and the extracted patterns could always be associated with the musical structure of the piece. For other pieces, some patterns could be found that do not begin or end at the right temporal position in the sequence. Together with this issue, the consideration of all the notes of the polyphonic context was sometimes a constraint for recognizing the repetition of the only melodic line (for instance in canons pieces). The relations between the two (polyphony and the melodic line) is one of the main issue of pattern extraction.

7 Conclusion

We have presented a general system for pattern extraction from polyphonic musical sequences. The notion of perceptible pattern has been discussed in the context of polyphonic music. A global similarity measure that takes into account rhythm, pitches and contour in a polyphonic context has been proposed. A method for extracting patterns from similarity matrices has been described and we have provided several musical examples for illustration. The method we propose could be used in several different musical applications (extraction of some components of musical structure, motive discovery, characterization of different musical pieces from their rhythmical or pitch components, search of allusions to a given pattern in a database of Midifiles). The similarity measure could also be adapted for applications such as "query by constraints": instead of humming a pattern, one could specify it with constraints on pitches or rhythm. The algorithm could then extract several patterns from different musical pieces with same rhythmical or pitch profile.

In future work, we plan to integrate temporal context in the model, in such a way that both polyphonic and temporal context are taken into account in the computation of the similarity matrices. We also plan to consider all the events of the sequence in the similarity measure and not only the events whose onsets correspond to the downbeat. Sequences with very different rhythm could then be compared and the quantifying step would not be necessary anymore. A prototype model has already been implemented. The only modification that we have made to the model concerns the algorithm that selects the sequences to compare. The similarity algorithm was not modified.

We think that our model considers interesting polyphonic aspects of music that are rarely taken into account, but several other issues remain to be solved, particularly the extraction of relevant patterns from the similarity matrix. However, we believe that solutions should emerge from such experimentations that come and go between perceptive considerations and modelization.

Acknowledgements

Part of this work was conducted in the context of the European project CUIDADO. Thank you to Olivier Lartillot for the interesting discussions that we have had on this subject.

References

1. Peteers G et al, *Toward Automatic Music Audio Summary Generation from Signal Analysis,* Proc Ismir, Paris, 2002,
2. Lartillot O, Perception-Based Advanced Description of Abstract Musical Content,*Proc. WIAMIS, London,* 2003

3. Rolland P-Y, Discovering patterns in musical sequences, *JNMR 28 N°4 pp. 334-350,* 1999
4. Meredith et all, *Algorithms for discovering repeated patterns in multidimensional representations of polyphonic music,* Cambridge Music Processing Colloquium, 2003
5. Cambouropoulos E, *owards a General Computational Theory of Musical Structure,*PhD, Edinburgh, Faculty of Music and Department of Artificial Intelligence, 1998
6. Meudic B, *A causal algorithm for beat tracking,* 2nd conference on understanding and creating music, Caserta, Italy, 2002
7. Tversky A, *Features of similarity,*ournal of Psychological Review, p327-352 1977

Deriving Musical Structures from Signal Analysis for Music Audio Summary Generation: "Sequence" and "State" Approach

Geoffroy Peeters

Ircam
1, pl. Igor Stravinsky, 75004 Paris, France
peeters@ircam.fr
http://www.ircam.fr

Abstract. In this paper, we investigate the derivation of musical structures directly from signal analysis with the aim of generating visual and audio summaries. From the audio signal, we first derive features - static features (MFCC, chromagram) or proposed dynamic features. Two approaches are then studied in order to derive automatically the structure of a piece of music. The sequence approach considers the audio signal as a repetition of sequences of events. Sequences are derived from the similarity matrix of the features by a proposed algorithm based on a 2D structuring filter and pattern matching. The state approach considers the audio signal as a succession of states. Since human segmentation and grouping performs better upon subsequent hearings, this natural approach is followed here using a proposed multi-pass approach combining time segmentation and unsupervised learning methods. Both sequence and state representations are used for the creation of an audio summary using various techniques.

1 Introduction

Analysis of digital media content, such as digital music, has become one of the major research fields in the past years, given the increasing amount and the increasing facilities for accessing large sets of them. Music identification (Audio-ID), music search by acoustic similarities and music structure discovery are among the new research areas of what is called Music Information Retrieval. Automatic music structure discovery from signal analysis aims specifically at satisfying demands for accessing and browsing quickly through digital music catalogs by content. Among its potential applications are:

- *Browsing/accessing music catalogs* by specific inner-keys (browsing through the chorus or theme of a music catalog),
- *Browsing through music items* (browsing a music title forward or backward by verse/ chorus/ solo),
- *Audio summary (or audio thumbnails)*: creating a short audio file representing the content of a piece of music[1],

[1] Note that not all audio summary techniques use the music structure in order to create an audio summary.

U.K. Wiil (Ed.): CMMR 2003, LNCS 2771, pp. 143–166, 2004.

- *Educational*: giving a quick and easily understandable (without requiring musical education) insight into temporal organization of a piece of music,
- *Musicology*: helping musicologists analyze pieces of music through the analysis of performances (rather than scores) and allowing their comparison,
- *Creativity*: re-mixing already existing pieces of music using structure[2].

Discovering the musical structure of a piece of music has been based for a long time on the analysis of symbolic representations (such as notes, chords, rhythm) [6] [17] [8]. It was therefore obvious to attempt deriving this symbolic representation directly from the signal [9] (pitch estimation, multiple pitch-estimation, beat tracking). Considering the weak results currently obtained for the task of deriving symbolic representation from the signal, other approaches have been envisioned. Approaches, like score alignment [22], attempt to link a given symbolic representation (MIDI file) to a digital signal. The signal therefore benefits from all the analysis achievable on a symbolic representation. However, since a symbolic representation is most of the time unavailable and since its derivation from the signal is still difficult to achieve, people started thinking of deriving directly the structure from lower-level signal features.

Music structure discovery from signal analysis takes its sources back from the works on signal segmentation first developed for speech applications and later used for musical applications. The question was then "what does the actual segments of the music represent ?".

1.1 Music Structure Discovery

Similarity Definition: All the techniques proposed so far for music structure discovery from signal analysis are based on a search for repetitions of motives or of melodies. This kind of approach is, of course, only applicable to certain kinds of musical genres based on some kind of repetition. The search of repetitions is based on measuring the similarity between signal observations or groups of observations. This similarity is then used in order to group the observations two by two (or into clusters) or oppositely to segment the temporal flows of observations into segments. The similarity measure is based on the definition of a distance between two observations (or two groups of observations): Euclidean distance, scalar product, cosine distance, normalized correlation, symmetric Kullback-Leibler distance or the Mahalanobis distance.

Similarity Matrix: The similarity matrix was first proposed by [11] under the name "Recurrence Plots". It was latter used in [13] for music structures discovery. If we note $s(t_1, t_2)$ the similarity between the observations at two instants t_1 and t_2, the similarity of the feature vectors over the whole piece of music is defined as a similarity matrix $\underline{\underline{S}} = |s(t_i, t_j)|\ i, j = 1, \dots, I$. Since the distance is symmetric, the similarity matrix is also symmetric. If a specific segment of music ranging

[2] This remix process is comparable to the re-arrangement of drum-loops allowed in Steinberg© Recycle software which allows people to recompose new drum loop patterns from small slides resulting from the automatic segmentation of the drum loops into rhythmical elements.

from times t_1 to t_2 is repeated later in the music from t_3 to t_4, the succession of feature vectors in $[t_1, t_2]$ is supposed to be identical (close to) the ones in $[t_3, t_4]$. This is represented visually by a lower (upper) *diagonal* in the similarity matrix.

Signal Observations: The observations derived from the signal, used to compute the similarity, play an essential role in the obtained results. Various types of features have been proposed for the task of music structure discovery:

- *Mel Frequency Cepstral Coefficients (MFCCs):* the MFCCs have been first proposed in the Automatic Speech Recognition (ASR) community [15]. The MFCCs are derived from the DCT of the logarithmic spectrum previously grouped into Mel bands (logarithmically spaced frequency bands). The MFCCs can be viewed as a sinusoidal decomposition of the Mel spectrum and allows representing its global shape with only a few coefficients (usually 12-13 coefficients).
- *Chromagram:* the chromagram was proposed by [3] in order to represent the harmonic content of the spectrum. The chromagram is estimated by grouping the logarithmic spectrum bins according to their pitch-class, whatever their octave-class is. For a western musical scale, 12 coefficients are used.
- *Scalar features:* scalar features derived from the waveform or from the spectrum such as the spectral centroid, spectral rolloff, spectral flux, zero-crossing rate, RMS or Energy are often used for the task of signal segmentation [28] [14] [29] [25]
- *Mean and standard deviation of MFCCs:* in order to reduce the amount of observations needed to represent a whole piece of music, [29] propose to summarize the observation content over a short period of time by taking the statistical moment of the observations.

In part 2, we present new types of features called dynamic features

1.2 Sequence and State Approach

Whatever distance and features used for observing the signal, music structure discovery techniques can be mainly divided into two types of approaches:

- the "sequence" approach: which consider the music audio signal as a repetition of sequences of events. These methods rely mainly on the analysis of the similarity matrix.
- the "state" approach: which consider the music audio signal as a succession of states. These methods rely mainly on clustering algorithms.

1.3 Related Works

"Sequence" Approach: Most of the sequence approaches start from Foote's works on the *similarity matrix*. Foote showed in [13] that a similarity matrix applied to well-chosen features allows a visual representation of the structural information of a piece of music, especially the detection of a sequence of repetitions through the lower (upper) diagonals of the matrix. The signal's features used in his study are the MFCCs.

The similarity matrix can be used for the determination of the *direct location of the key sequence* of a piece of music used then as the audio summary. In [3], a similarity matrix is computed using the chromagram parameterization. A uniform average filter is then used in order to smooth the matrix. The maximum element of the matrix, supposed to be the most representative, is then chosen as the audio summary. In [7], a summary score is defined using the "average of the MFCCs' similarity matrix rows" over a specific interval. For a specific interval size, the starting time of the interval having the highest summary score is chosen as the starting time of the audio summary.

The similarity matrix can also be used for *discovering the underlying structure* of a piece of music. In [2], two methods are proposed for diagonal line detection in the MFCCs' similarity matrix: a "Gaussian distribution extruded over the diagonal with a reinforced diagonal" filter, and a computationally more expensive process: the "Hough Transform". Pattern matching techniques (including deletion, insertion and substitution processes) are then used in order to derive the structure from the detected diagonals. In [9], several algorithms based on dynamic programming are proposed in order to derive the structure using either monophonic pitch estimation, polyphonic transcription or chromagram features.

In part 3, we present a novel method for line detection in the lag-matrix based on a 2D structuring filter. It is combined with a pattern matching algorithm allowing the description of the music in terms of sequence repetitions.

"State" Approach: The similarity matrix can be used in order to perform simply (i.e. without going into structure description) the *segmentation* of a piece of music. In [12], a measure of novelty is proposed in order to perform music segmentation into long structures (such as chorus/verse), or into short structures (such as rhythmic structures). This measure of novelty is computed by applying a "checkerboard" kernel to the MFCCs' similarity matrix. The novelty is defined by the values of the convoluted similarity matrix over the main diagonal.

Unsupervised algorithms are most of the time used in order to obtain the *state representation.* A study from Compaq [18] uses the MFCC parameterization in order to create "key-phrases". In this study, the search is not for lower (upper) diagonal (sequence of events) but for states (collection of similar and contiguous states). The song is first divided into fixed length segments, which are then grouped according to a cross-entropy measure. The longest example of the most frequent episode constitutes the "key-phrase" used for the audio summary. Another method proposed by [18] is based on the direct use of a hidden Markov model applied to the MFCCs. While temporal and contiguity notions are present in this last method, poor results are reported by the authors. In [1], a Hidden Markov Model (with gaussian mixture observation probabilities) is also used for the task of music segmentation into states. Several features are tested: MFCC, Linear Prediction and Discrete Cepstrum. The authors conclude on better results obtained using the MFCCs and reports the limitation of the method due to the necessity to fix a priori the number of states of the HMM.

In part 4, we present a novel multi-pass algorithm combining segmentation and HMM which does not require the a priori fixing of the number of states.

1.4 Audio Summary or Audio Thumbnailing

Music summary generation is a recent topic of interest. As a significant factor resulting from this interest, the recent MPEG-7 standard (Multimedia Content Description Interface) [20], proposes a set of meta-data in order to store multimedia summaries: the Summary Description Scheme (DS). This Summary DS provides a complete set of tools allowing the storage of either sequential or hierarchical summaries.

However, while the storage of audio summaries has been normalized, few techniques exist allowing their automatic generation. This is in contrast with video and text where numerous methods and approaches exist for the automatic summary generation. Without any knowledge of the audio content, the usual strategy is to take a random excerpt from the music signal, or an excerpt in the middle of it. In speech, time-compressed signals, or time-skipped signals are preferred in order to preserve the message. A similar strategy can be applied in music by providing excerpts from meaningful parts of the music derived from its structure.

In part 5, we present a novel approach for audio summary generation based on choosing specific excerpt of the signal derived from the sequence/state representation.

2 Signal Observation: Dynamic Audio Features

Signal features, such as MFCCs, chromagram, ... are computed on a frame by frame basis, usually every 10 ms. Each of these features represents a specific description (description of the spectral shape, of the harmonic content, ...) of the signal *at (around) a given time.* Since the amount of features obtained can be very large (for a 4 minutes piece of music, 4*60*100= 24000 feature vectors), and thus hard to process by computer (the corresponding similarity matrix is 24000*24000) and hard to understand, the features are often "down-sampled" by use of mean (low-pass filter), standard deviation or derivative values (high-pass filter) over a short period of time. However this kind of sampling is not adequate to represent the temporal behavior of the features (for example to represent a regular modulation). Modeling the temporal evolution along time is the goal of the *"dynamic"* features which are proposed here. On the opposite, we call features, such as MFCCs, chromagram, ... *"static"* features since they don't represent any temporal evolution of a description.

In dynamic features, the evolution of a feature along time is modeled with a Short Time Fourier Transform (STFT) applied to the signal formed by the values of the feature along time (in usual signal processing, the STFT is applied to the audio signal): around each instant t, the time evolution of the feature on a specific duration L is modeled by its Fourier Transform. If the feature is multi-dimensional (as it is the case of the MFCCs), the Fourier Transform is applied to each dimension of the feature.

Among the various types of possible dynamic features, the best results were obtained by modeling the time evolution of the energy output of an *auditory filterbank.* The audio signal $x(t)$ is first passed through a bank of N Mel filters.

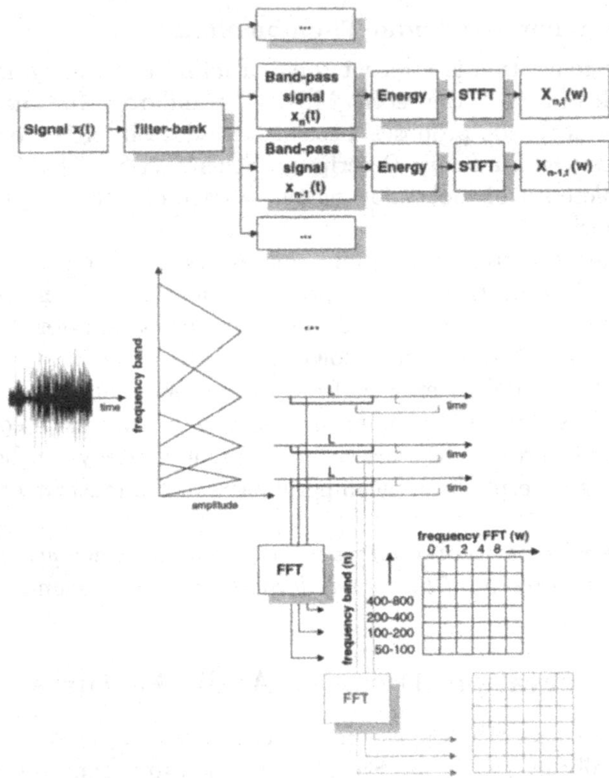

Fig. 1. Dynamic features extraction process. From left to right: the signal is passed through a filterbank; the energy over time of the output signals of each filter (band-pass signal) is estimated; each signals is modeled with a STFT of size L; at each time, the resulting FFT are grouped in a matrix representing the amplitude at each frequency of the FFT (w) for each frequency band (n).

The slow evolution ([0-50] Hz) of the energy of each output signal $x_n(t)$ of the $n \in N$ filters is then analyzed by Short Time Fourier Transform (STFT). Since the frequencies we are interested in are low, the hop size of the STFT can be large (up to 1s). The output of this is, at each instant t, a matrix $X_{n,t}(\omega)$ representing the amplitude of the variations at several speeds ω of several frequency bands n observed with a window of size L. The feature extraction process is represented in Fig. 1.

Dynamic features can represent slow variations (small ω) at low frequencies (small n) as well as quick variations (large ω) at high frequencies (large n) or any other combination between speed of variation and frequency band.

Another important parameter of dynamic features is the window size L on which the modeling is performed (the length of the analysis window used for the STFT analysis). Using large window tends to favor a "state" representation of the music. Since the modeling is performed on a long duration, the resulting spectrum of the FFT represents properties which are common to all features evolution on a long duration. Therefore the modeling tends to catch the arrangement part of the music. On the opposite, using short window tends to favor the "sequence" representation.

When using dynamic features, only the most appropriate combination of several frequency bands n in several speeds of variation ω for a specific application is kept. This can be done for example by using the first Principal Components of a PCA as has been proposed by [27].

In Fig. 2, Fig. 3, Fig. 4 and Fig. 5, we compare the similarity matrices obtained using MFCCs, dynamic features with short, middle and long duration of the model for the title "Natural Blues" by "Moby" [19].
The title "Natural Blues" by "Moby" is used several times in this paper in order to illustrate our studies. This title has been chosen for several reasons: 1) the structure of the arrangement of the music (the instruments playing in the background) does not follow the structure of the melodic part (usually in popular music both arrangement and melody are correlated). Therefore it provides a good illustration of the difference between the sequence and the state representation; 2) the various melodic parts are all based on sampled vocals. Therefore there is an exact sequence repetitions (usually in popular music, when a chorus is repeated it can vary from one repetition to the other, therefore the question appears "is it the same melody, a variation of it or a different one ?").
In Fig. 2, the similarity matrix using MFCCs is represented for the first 100 s of the music. We see the repetition of the sequence 1 "oh lordy" $t = [0, 18]$ *at* $t = [18, 36]$*, the same is true for the sequence 2 "went down"* $t = [53, 62]$ *which is repeated at* $t = [62, 71]$*. In Fig. 3, the similarity matrix using dynamic features with a short duration (L=2.56s) is represented for the whole title duration (252 s). Compared to the results of Fig. 2, we see that the sequence 1* $t = [0, 18]$ *is in fact not only repeated at* $t = [18, 36]$ *but also at* $t = [36, 54]$, $t = [72, 89]$, $[89, 107]$, $[160, 178]$, *... This was not visible using MFCC parameterization because the arrangement of the music changes at time* $t = 36$ *masking the sequence repetition. Note that the features' sampling rate used here is only 4 Hz (compared to 100 Hz for the MFCC). In Fig. 4 and Fig. 5 we compare middle and long duration of the model (L=5.12s and L=10.24s) for the representation of the structure of the arrangement of the song. Fig. 4 and Fig. 5 shows the introduction at* $t = [0, 36]$*, the entrance of the first rhythm at* $t = [36, 72]$*, the main rhythm at* $t = [72, 160]$*, the repetition of the introduction at* $t = [160, 196]$*, the repetition of the main rhythm at* $t = [196, 235]$*, and ending with a third repetition of the introduction at* $t = [235, 252]$*. Fig. 4 shows a more detailed description of the internal structure of each part. Note that the features sampling rate used here is only 2 Hz for Fig. 4 and only 1 Hz for Fig. 5.*

3 Sequence Approach

A high value in the similarity matrix $\underline{\underline{S}}(t_i, t_j)$ represents a high similarity of the observations at times t_i and t_j. If a sequence of events at time $t_i, t_{i+1}, t_{i+2}, \ldots$ is similar to another sequence of events at time $t_j, t_{j+1}, t_{j+2}, \ldots$ we observe a lower (upper) diagonal in the matrix.

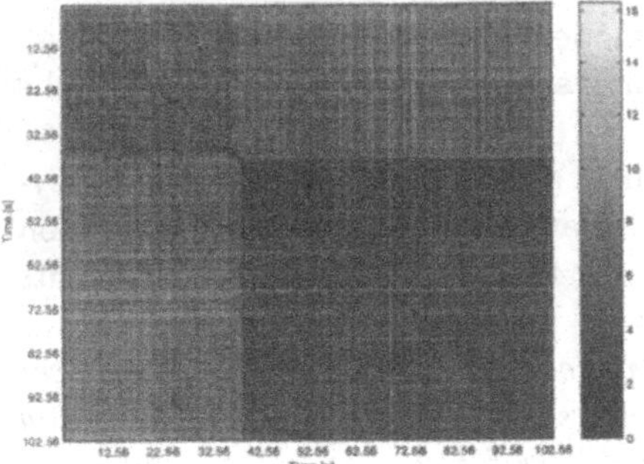

Fig. 2. Similarity matrix using MFCC features for the title "Natural Blues" by "Moby".

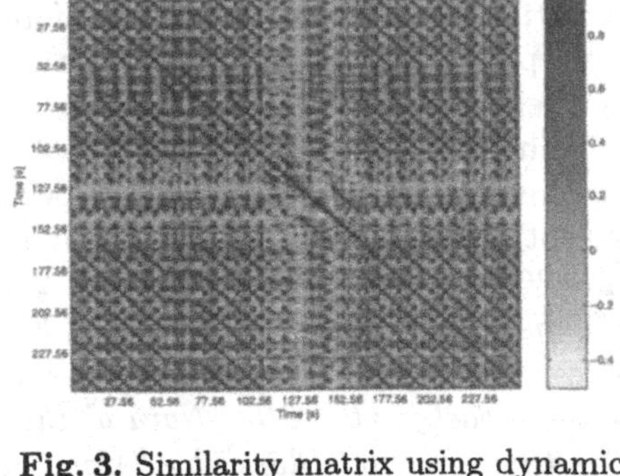

Fig. 3. Similarity matrix using dynamic features with short duration modeling (L=2.56s).

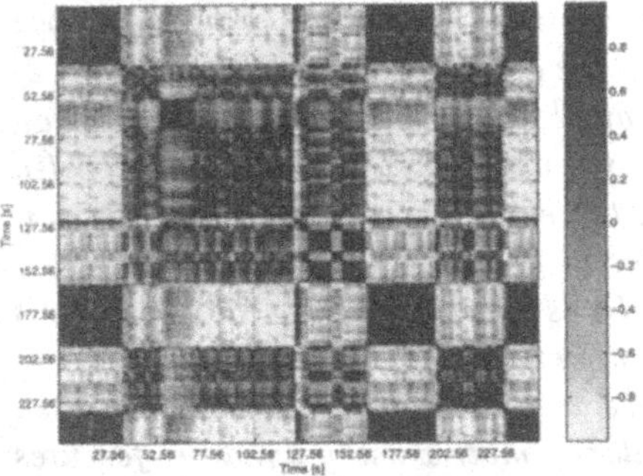

Fig. 4. Similarity matrix using dynamic features with middle duration modeling (L=5.12s).

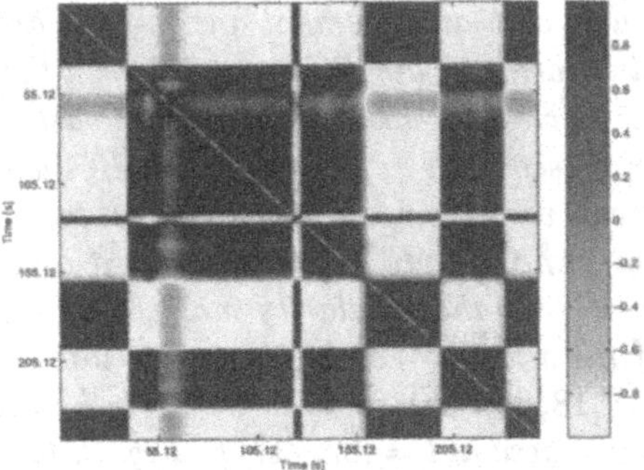

Fig. 5. Similarity matrix using dynamic features with long duration modeling (L=10.24s).

The lag between the repetition (starting at t_i) and the original sequences (starting at t_j) is given by projecting t_i on the diagonal of the matrix and is therefore given by $t_i - t_j$. This is represented in the lag-matrix $\underline{\underline{L}}$: $\underline{\underline{L}}(t_i, lag_{ij}) = \underline{\underline{S}}(t_i, t_i - t_j)$. The diagonal-sequences in the similarity-matrix become line-sequences in the lag-matrix. This representation is used here since processing on lines is easier.

As an example lag matrix of the title "Natural Blues" by "Moby" is represented in Fig. 6.

There exist many articles about the similarity-matrix or lag-matrix but few of them address the problem of deriving from this visual representation the actual time pointers to the sequence's repetitions. This is the goal of this section.

Our approach for deriving a sequence representation of a piece of music works in three stages:

1. from the feature similarity/lag matrix we first derive a set of *lines* (a line is defined here as a possibly discontinuous set of points in the matrix) (part 3.1)

2. from the set of lines we then form a set of *segments* (a segment is defined here as a set of continuous times). Since we only observe repetitions in the similarity/lag matrix, a segment derived from a line is a repetition of some original segment. A segment is therefore defined by the staring time, the ending time of its repetition and a lag to its original segment (part 3.2).
3. from the set of segments (original and repetition segments) we finally derive a *sequence* representation (a sequence is defined by a number and a set of time intervals where the sequence occurs; a sequence representation is defined by a set of sequences) (part 3.3).

The global flowchart of the proposed algorithm for sequence representation is represented in Fig. 18.

3.1 Search for Lines in the Matrix

In order to facilitate the detection of line (or diagonal) in the matrix, usually people [3] [2] first apply 2D filtering to the matrix in order to increase the contrast between sequences and the so-called "noisy" similarity. For the detection of vertical lines in a lag-matrix, a vertical low-pass and horizontal high-pass filter can be applied. However, use of kernel based filtering techniques, if it allows one to get rid of most of the "noisy" similarity, blurs the values and therefore prevents the detection of the exact start and end positions of a segment. Moreover, it constitutes a step toward the detection of start and end positions of a segment but does not give them (one still needs to find where the sequence starts in the continuous set of values of the matrix). For this reason, we studied the applicability of a 2D structuring filter.

Structuring Filters: For a specific data point y, structuring filters use neighboring values $[y - layy, y + lagy]$ to decide on keeping the value of y or canceling it. This choice is based on the local mean around y. This can be expressed in a MATLAB© way as:

```
if y < mean([y-lagy:y+lagy])
then y=0
else y=y
```

The 2D structuring filter method we propose for vertical lines detection (see Fig. 12) is based on counting the number of values in the neighboring interval $[y-lagy, y+lagy]$ which are above a specific threshold $t1$. If this number is below another threshold $t2$ then y is canceled. This can be expressed in a MATLAB© way as:

```
if y < length( find([y-lagy:y+lagy]) >t1 ) < t2
then y=0
else y=y
```

The first threshold, $t1$, allows one to get rid off the low values in the similarity matrix. The second threshold, $t2$, is proportional to the size of the considered interval: $t2 = k * (2 * lagy + 1)$, where k ranges from 0 (no values need to be above $t1$) to 1 (all values must be above $t1$).

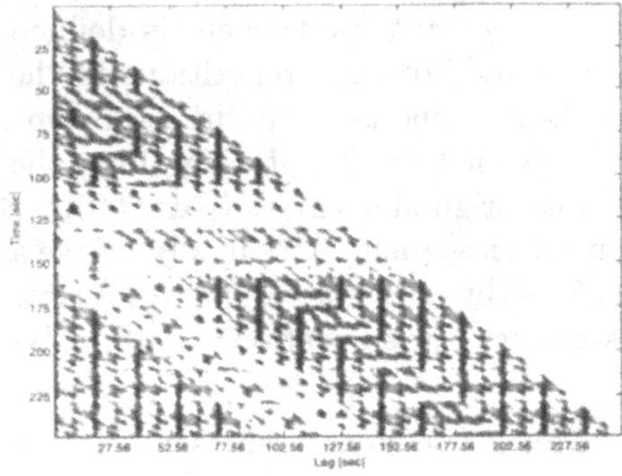

Fig. 6. Lag-matrix (X=lag, Y=time) for the title "Natural Blues" by "Moby".

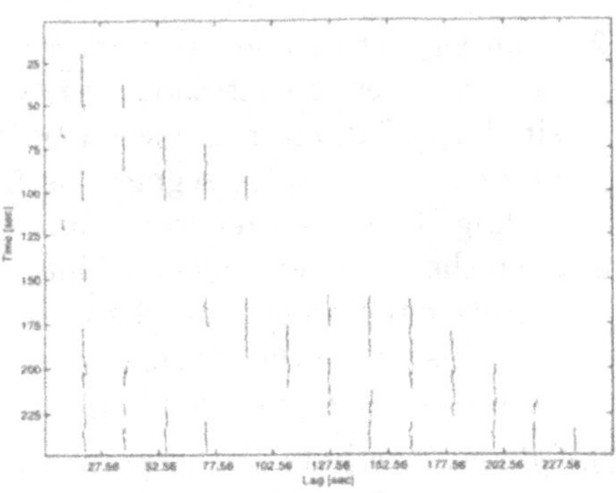

Fig. 7. Result of applying the 2D structuring filter to the lag matrix (X=lag, Y=time).

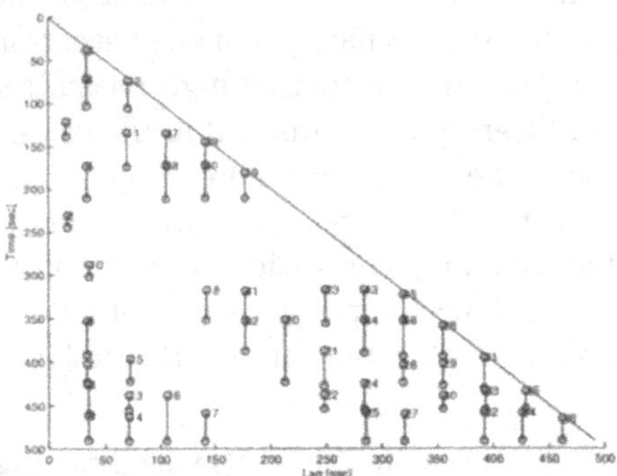

Fig. 8. Detected segments along time (X=lag, Y=time).

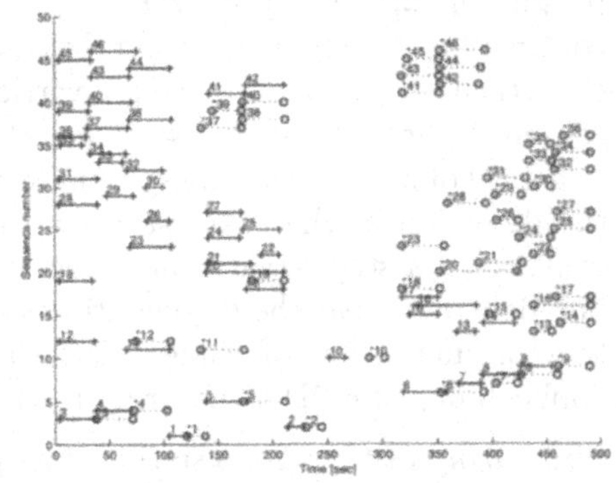

Fig. 9. Original segments (–) and repeated segments (...) along time (X=time, Y=segment number).

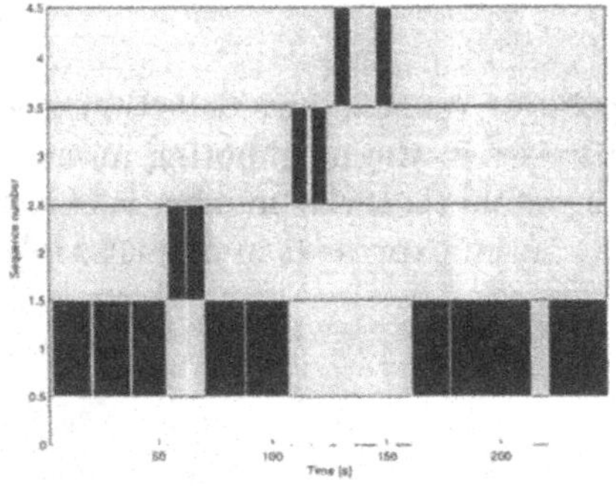

Fig. 10. Detected sequence repetitions along time (X=time, Y=sequence number).

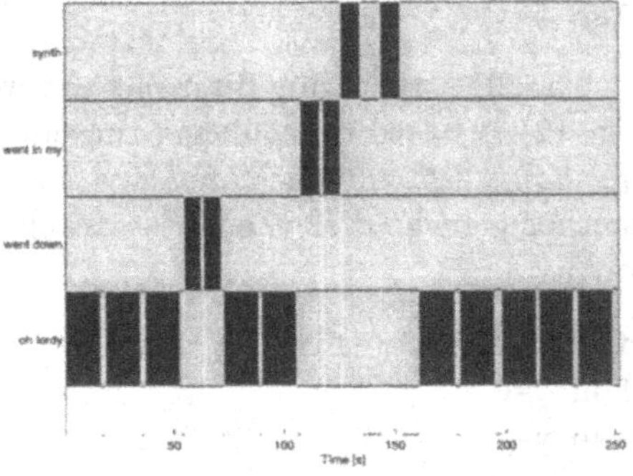

Fig. 11. Real sequence repetitions along time (X=time, Y=sequence number).

Since a sequence can be repeated at a slower or quicker rate (resulting in a departure of the line-sequence from the column x to its neighboring column $x - lagx$ or $x + lagx$), we extend the 2D structuring filter in order to take also into account the contribution of the neighboring columns $[x - lagx, x + lagx]$: if, for a specific y, at least one of the values on the interval $([x - lagx, x + lagx], y)$ is above a specific threshold $t1$ then a new hit is counted (there is no cumulative total across y).

In order to avoid that all the contribution to the counter would come from a neighboring column, we add the condition that the main contribution to the counter must come from the main column x.

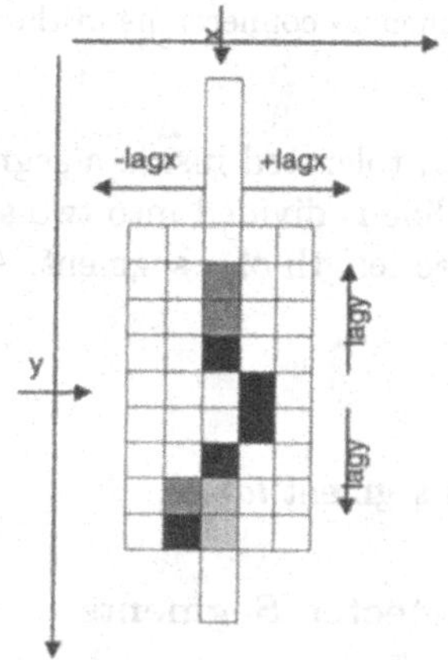

Fig. 12. Two dimensional structuring filter for vertical lines detection.

The result of the application of the proposed 2D structuring filter on the lag-matrix of Fig. 6 is represented on Fig. 7.

Avoiding doubled lines: Resulting from the previous stage is a subset of lines detected in the lag-matrix. However, because of the extension to the consideration of the neighboring columns x in our algorithm[3], doubled lines detection is possible. For an analysis hop size of 0.5 s, and by the definition of a lag-matrix, two neighboring lines represent two repetitions of the same sequence separated by 0.5 s. We remove doubled lines by defining the minimum delay between two sequences' repetition (fixed to a value of 5 s in our experiment - which means that we won't be able to detect a sequence repeated with a period less than 5 s). When several sequences are neighboring, only the longest one is kept.

3.2 From Lines to Segments

Detected lines can be discontinuous (the line suddenly disappears during some values of y). In order to be able to define segments, we need to define the max-

[3] Depending on the value of $lagx$, a specific value (x, y) can be shared by several columns.

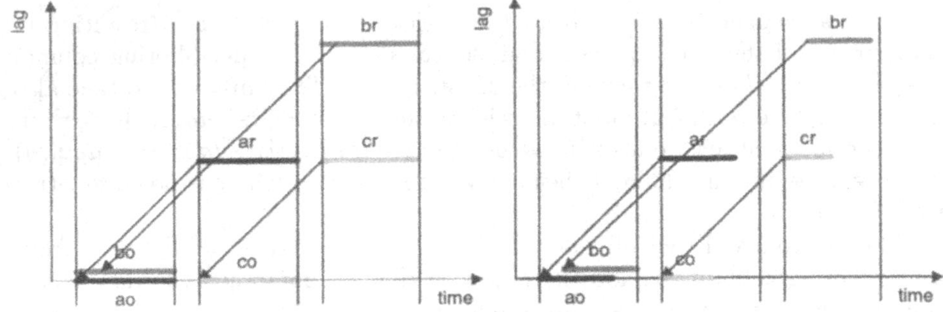

Fig. 13. Sequences connections in the lag-matrix.

imum length of a gap, $G_{\max}$, tolerated inside a segment. If the observed gap is larger than $G_{\max}$, then the line is divided into two separated segments. We also define the minimum accepted length of a segment. A detected segment i is now defined by

- its start time $s(i)$,
- its end time $e(i)$,
- its lag from the original segment $lag(i)$

3.3 Interpreting the Detected Segments

Connection between the various segments involves 1) finding which segment is a repetition of which other one 2) finding which segment can be considered as the reference sequence.

Each segment i is in fact defined by its repetition segment ir (the detected line) and its original segment io (the detected line translated by lag). In the left part of Fig. 13, we represent the ideal case. The segment a is defined by its repetition ar and its original segment ao. The same is true for b. We see that bo shares the same period of time as ao. a and b are therefore supposed to be identical sequences. This is verified by the presence of the segment c: since ao and bo are the same and ar and ao are the same, there must be a segment c which repetition cr shares the same period of time as br and which original co shares the same period of time as ar. However what we observe in practice is closer to the right part of Fig. 13: bo and ao only share a portion of time (hence the question "which one is the reference ?"); the segment c can be very short, co is in ar but, since it shares a too short period of time with br, it is not in br. This is in contradiction with the transition rule: $cr \to co \to ar \to ao \to bo \to br$.

*A real case example is represented in Fig. 8 and Fig. 9 for the same signal as Fig. 6. Fig. 8 represents the detected segments in the lag/time space. Starts and ends of segments are indicated by circles (o). Fig. 9 represents each segments original (–) and repetition (...) along time. X-axis represents the time, Y-axis represents the segment's number. In this case, we need to connect 92 (2*46) segments with each other.*

Proposed algorithm for segments connection: In order to perform the connection between segments, and form sequences, the following algorithm is proposed. Two segments are said to belong to the same sequence if the period of time shared by the two segments is larger than a specific amount of their own duration (we have chosen a value of 70%). If they share less than this amount they are said to be different. The algorithm works by processing segments one by one and adding them to a sequence container. We note

- jO the original of a new segment and jR the repetition of a new segment (the detected line)
- I the sequence container
- iO the original of a segment already present in the sequence container.

```
Init: define min_shared_time=0.7
Init: add first jO and jR to I
_while there is non-processed segment j
__take a new segment j (original jO of length jOL)
__if new segment jO shares time with a iO in the
container
___for each of these i,
        define iOL = length of iO,
        define jOL = length of jO,
        define ijOL= the shared time length of iO and jO,
        define c1  = ratios c1=ijOL/jOL,
        define c2  = ratios c2=ijOL/iOL
____select the i with the largest c1+c2
____if c1 > min_shared_time | c2 > min_shared_time then repetition
_____if c1 > min_shared_time & c2 < min_shared_time then jO is in iO
_____if c1 < min_shared_time & c2 > min_shared_time then iO is in jO
_____add jR to I with the same sequence tag as iO
____else
_____add jO and jR to I with a new sequence tag
____end
__else
___add jO and jR to I with a new sequence tag
__end
_end
```

At this stage, I contains all the segments with an associated "sequence number" tag. The second stage of the algorithm decides, for each sequence number, which of its segments can be considered as the reference segment[4] and which of the segments initially supposed to be part of the sequence are in fact poorly explained by the other segments of this sequence. In order to do that, for each sequence number, each of its segments is in turn considered as the reference segment of the sequence. For each candidate reference segment, we estimate how

[4] The reference segment of a sequence is the one that best explains the other segments of the sequence.

many and how much of the other segments can be explained by using this reference segment. This is done by computing a score defined as the amount of time shared by the candidate reference segment - original and repetition - and all the other segments[5]. The reference segment is the candidate segment with the highest score. The remaining segments that were highly explained by the reference segment are attributed to this sequence. Finally, the reference segment and the attributed segments are removed from the container I and the process is repeated as long as there are still non attributed segments in I.

3.4 Results

The result of the application of our algorithm of segment connection to the detected segments of Fig. 8 and Fig. 9 is illustrated in Fig. 10 (title "Natural blues" by "Moby"). For comparison, the real structure (manually indexed) is represented in Fig. 11. Three different sequences were detected. Sequence 1 is the "oh lordy, trouble so hard" melody. Sequence 2 is the "went down the hill" melody. Sequence 3 is the "went in the room" melody. Sequence 4 is a melody played by the synthesizers appearing twice. Compared to the real sequence structure of the song, we can see that two occurrences of sequence 1 have not been detected in the last part of the song.

We also illustrate our sequence representation method in Fig. 14, Fig. 15, Fig. 16 and Fig. 17 for title "Love me do" by "The Beatles" [4]. Fig. 14 represents the lag matrix. The segments detected in the matrix using the 2D structuring filters are represented in Fig. 15. The resulting sequence representation is represented in Fig. 16. For comparison, the real structure (manually indexed) is represented in Fig. 17. Three different sequences were detected. Sequence 1 is the harmonica melody played several times across the song, sequence 2 is the "love me do" melody and sequence 3 is the "someone to love" melody. Note that the second occurrence of sequence 3 is in fact the same melody "someone to love" but played by the harmonica. The only false detection is the sequence 1 at time 450.

4 State Approach

The goal of the state representation is to represent a piece of music as a succession of states (possibly at different temporal scales) so that each state represents (somehow) similar information found in different parts of the piece.

The states we are looking for are of course specific for each piece of music. Therefore no supervised learning is possible. We therefore employ unsupervised learning algorithms to find out the states as classes. Several drawbacks of unsupervised learning algorithms must be considered:

[5] The score used is the same as in the previous code; it was noted $c1 + c2$.

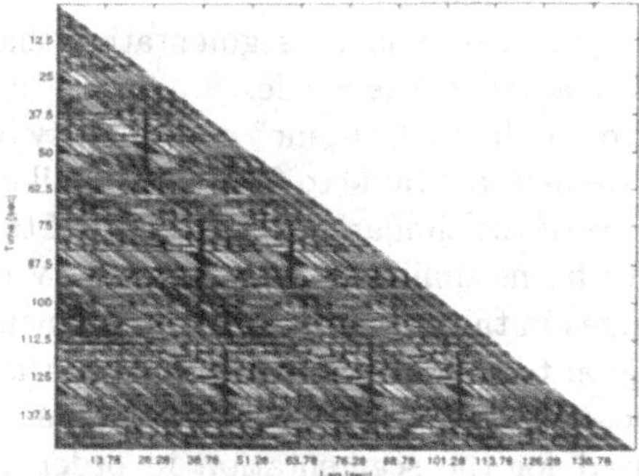

Fig. 14. Lag-matrix (X=lag, Y=time) for the title "Love me do" by "The Beatles".

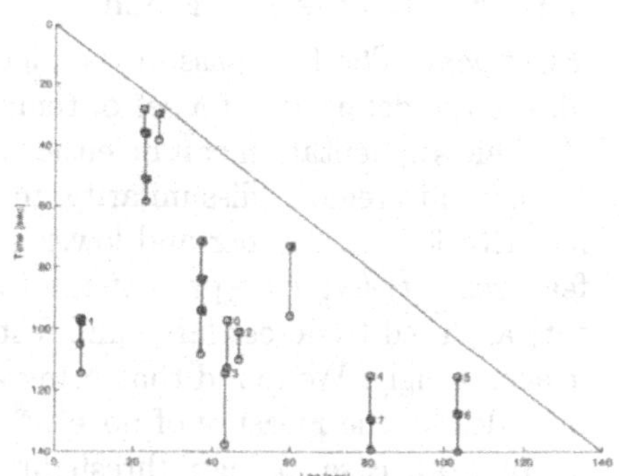

Fig. 15. Detected segments along time (X=lag, Y=time).

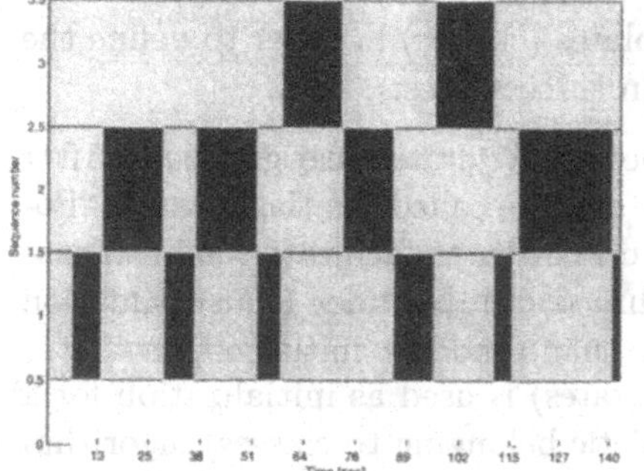

Fig. 16. Detected sequence repetitions along time (X=time, Y=segment number).

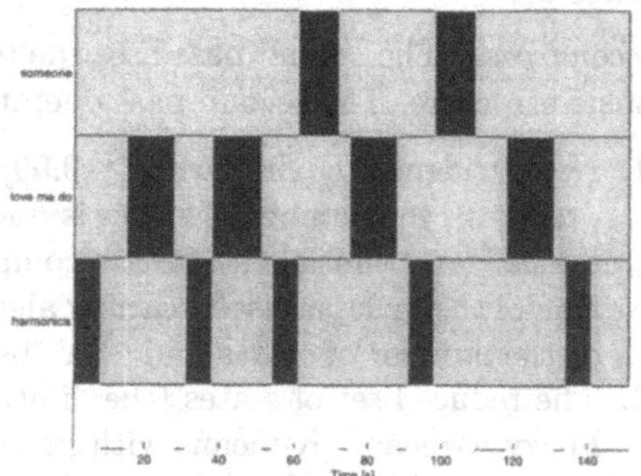

Fig. 17. Real sequence repetitions along time (X=time, Y=sequence number).

- usually a previous knowledge of the *number of classes* is required for these algorithms,
- these algorithms depend on a good *initialization of the classes*,
- most of the time, these algorithms do not take into account contiguity (spatial or temporal) of the observations.

A new trend in video summary is the *"multi-pass"* approach [31]. As for video, human segmentation and grouping performs better when listening (watching in video) to something for the second time [10]. The *first listening* allows the detection of variations in the music without knowing if a specific part will be repeated later. The *second listening* allows one to find the structure of the piece by using the previous mentally created templates. In [23] we proposed a multi-pass approach for music state representation, we review it here briefly. This multi-pass approach allows solving most of the unsupervised algorithm's problems.

The global flowchart of the multi-pass approach for state representation is represented in Fig. 19.

4.1 Multi-pass Approach

First pass: The first pass of the algorithm performs a signal segmentation that allows the definition of a set of templates (classes) of the music.

This segmentation can be either based on a "frame to frame" dissimilarity or a "region to region" dissimilarity. In [23], we used a "frame to frame" dissimilarity criterion. The upper and lower diagonals of the similarity matrix $\underline{\underline{S}}(t)$ of the features $\underline{f}(t)$ (which represent the frame to frame similarity of the features vector) are used to detect large and fast changes in the signal content and segment it accordingly. We found that using a "region to region" dissimilarity, such that provides by the measure of novelty[6] proposed by [12] also gives good results.

In both case, a high threshold[7] is used for the segmentation in order to reduce the "slow variation" effect and to ensure that all times inside a segment are highly similar.

We use the mean values of $\underline{f}(t)$ inside each segment to define "potential" states $\underline{s}_k$.

Second pass: The second pass uses the templates (classes) in order to define the music structure. The second pass operates in three stages:

1. Nearly identical (similarity ≥ 0.99) "potential" states are grouped. After grouping, the number of states is now K and are called "initial" states. "Potential" and "initial" states are computed in order to facilitate the initialization of the unsupervised learning algorithm since it provides 1) an estimation of the number of states and 2) a "better than random" initialization of it.
2. The reduced set of states (the "initial" states) is used as initialization for a Fuzzy K-means (K-means with probabilistic belonging to classes) algorithm (knowing the number of states and having a good initialization). We note $\underline{s}'_k$ the states' definition obtained at the end of the algorithm and call them "middle" states.
3. In order to take music specific nature into account (not just a set of events but a specific temporal succession of events), the output states of the Fuzzy K-means algorithm are used for the initialization of the learning of a Markov model. Since we only observe $\underline{f}(t)$ and not directly the states of the network, we are in the case of a hidden Markov model (HMM) [24]. A state k produces observations $\underline{f}(t)$ represented by a state observation probability $p(\underline{f}|k)$. The state observation probability $p(\underline{f}|k)$ is chosen as a gaussian pdf $g(\mu_k, \sigma_k)$. A state k is connected to other states j by state transition probabilities $p(k, j)$. Since no priori training on a labeled database is possible we are in the case of ergodic HMM. The *training* is initialized using the Fuzzy K-means "middle" states $\underline{s}'(k)$. The Baum-Welch algorithm is used in order to train the model. The outputs of the training are the state observation probabilities, the state transition probabilities and the initial state distribution.

[6] The measure of novelty is defined by the values over the main diagonal of the similarity matrix after convolution by a "checkerboard" kernel.

[7] The threshold is automatically determined by exhaustively trying all possible values of the threshold and comparing them to the number of segments obtained in each case.

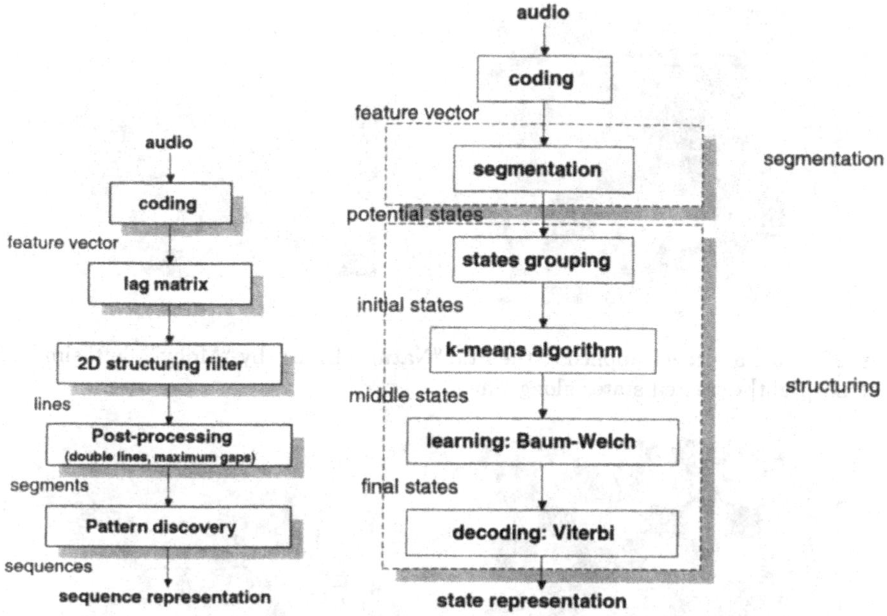

Fig. 18. Sequences representation flowchart.

Fig. 19. States representation flowchart.

4. Finally, the optimal representation of the piece of music as a HMM state sequence is obtained by *decoding* the model using the Viterbi algorithm given the signal feature vectors.

4.2 Results

The result of the proposed multi-pass approach is represented in Fig. 20, Fig. 21 and Fig. 24 for three different titles. The left parts of the figures shows the similarity matrix, the right parts of the figures shows the various states detected along time.

In Fig. 20, for the title "Natural Blues" by "Moby", five different states were found. Fig. 22 represents the "true" structure (manually indexed) of the title. Let us observe the correspondence between the detected states (right part of Fig. 20) and the label of the "true" structure (Fig. 22). State 1 represents the label [intro], state 2 is a mix between [drum1] and [drum2], state 3 is also a mix between [drum1] and [drum2]. Apparently, the algorithm didn't succeeded in catching the difference in the arrangements between [drum1] and [drum2] and rather focused on the variation of the melody (state 3 corresponds to the arrangements on "oh lordy", state 2 to the arrangements on the two other melodies). State 4 is the [synth] part which is correctly identified. State 5 is the [drum2] part which was detected as a separate state.

Fig. 21 represents the title "Smells Like Teen Spirit" by "Nirvana" [21]. Seven different states were found. Fig. 23 represents the "true" structure (manually

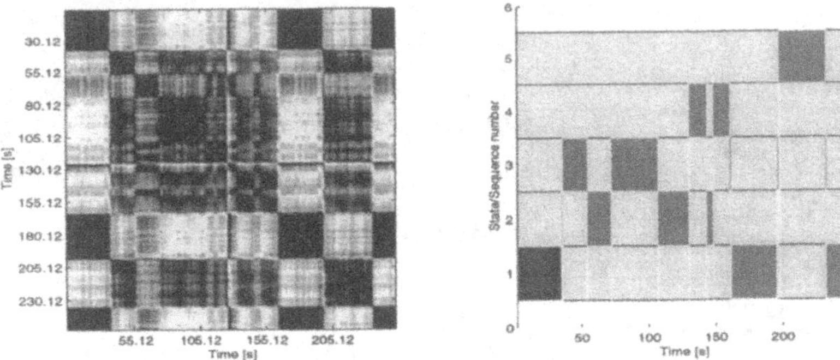

Fig. 20. State approach applied to the title "Natural Blues" by "Moby", [left] similarity matrix, [right] detected states along time.

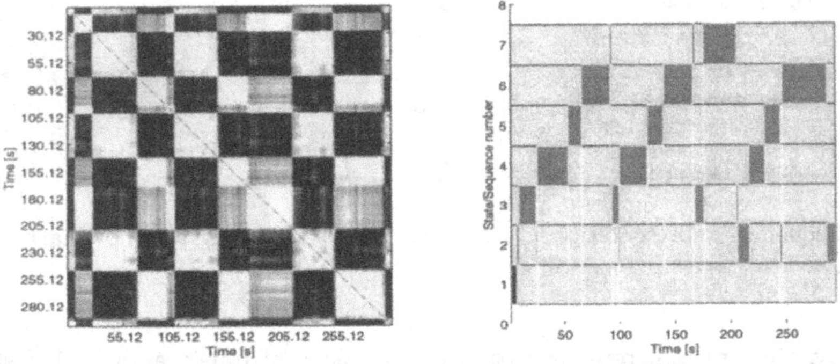

Fig. 21. State approach applied to the title "Smells like teen spirit" by "Nirvana", [left] similarity matrix, [right] detected states along time.

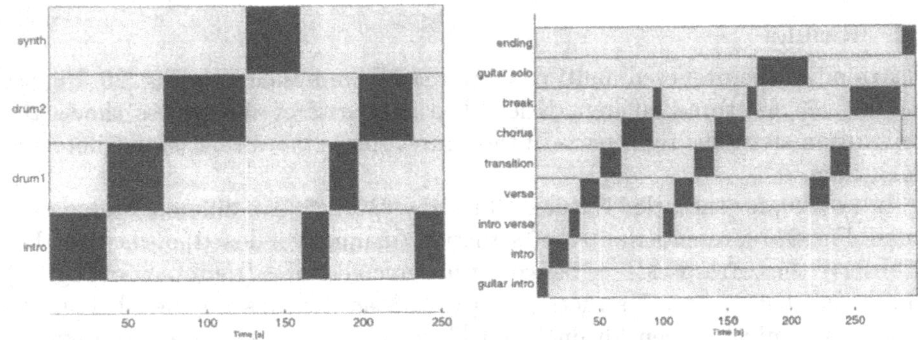

Fig. 22. "True" states along time of the title "Natural Blues" by "Moby".

Fig. 23. "True" states along time of the title "Smells like teen spirit" by "Nirvana".

indexed) of the title. Let us observe the correspondence between the detected states (right part of Fig. 21) and the label of the "true" structure [(Fig. 23). State 1 represents the label [guitar intro], state 2 seems to be a garbage state containing most drum rolls, state 3 contains the [intro], state 4 is the [verse],

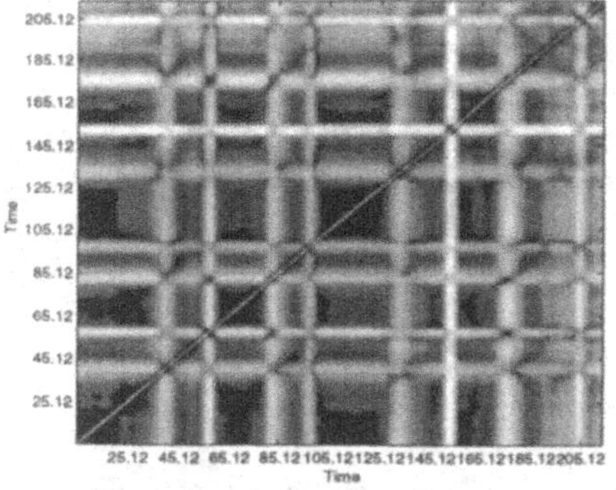

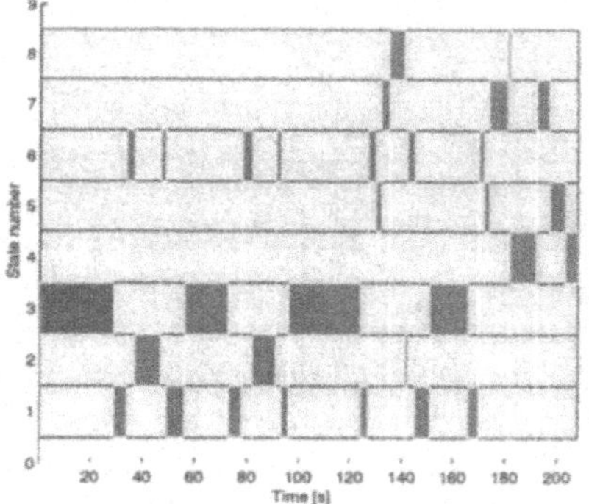

Fig. 24. State approach applied to the title "Oh so quiet" by "Bjork", [left] similarity matrix, [right] detected states along time.

state 5 the [transition], state 6 is the [chorus], state 7 represents both the [break] and the [guitar solo]. Considering that the most important part of the title is the (chorus/ transition/ verse/ solo), the detected representation is successful.

In Fig. 24 represents the title "Oh so quiet" by "Bjork" [5]. In this case the "true" structure of the title is hard to derive since most parts when repeated are repeated with large variations of the arrangement. Therefore, we show the obtained structure here only to check whether it makes sense. The characteristic verse/chorus repetition is very clear. State 3 represents the verse, state 1 the transition to the chorus, state 2 the chorus, state 6 the break, ...

5 Audio Summary Construction

So far, from the signal analysis we have derived features vectors used to assign a sequence number (through line detection in the similarity matrix) or a state number (through unsupervised learning) to each time frame. From this representation several possibilities can be taken in order to create an audio summary. Let us take as example the following structure: $AABABCAAB$. The generation of the audio summary from this sequence/state representation can be done in several ways (see Fig. 25):

- Each: providing a unique audio example of each sequence/state (A, B, C)
- All: reproducing the sequence/state successions by providing an audio example for each sequence/state apparition (A, B, A, B, C, A, B)
- Longest/most frequent: providing only an audio example of the most important sequence/state (in terms of global time extension or in term of number of occurrences of the sequence/state) (A)
- Transition: in the case of state representation: providing audio examples of state transitions $(A \to B, B \to A, B \to C, C \to A)$
- etc ...

This choice relies of course on user preferences but also on time constraints on the audio summary duration. In each case, the audio summary is generated by taking short fragments of the segment/state's signal. For the summary construction, it is obvious that "coherent" or "intelligent" reconstruction is essential.

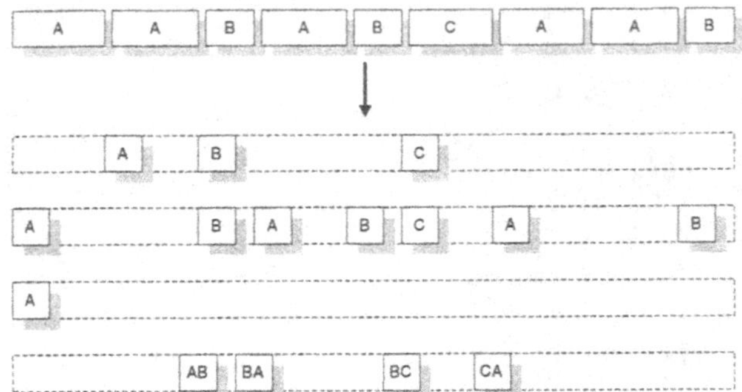

Fig. 25. Various possibilities for Audio Summary construction from sequence/state representation.

Information continuity will help listeners to get a good feeling and a good idea of a piece of music when hearing its summary:

- *Overlap-add:* The quality of the audio signal can be further improved by applying an overlap-add technique of the audio fragment.
- *Tempo/Beat:* For highly structured music, beat synchronized reconstruction allows improving largely the quality of the audio summary. This can be done 1) by choosing the size of the fragments as integer multiple of 4 or 3 bars, 2) by synchronizing the fragments according to the beat position in the signal. In order to do that, we have used the tempo detection and beat alignment proposed by [26].

The flowchart of the audio summary construction of our algorithm is represented in Fig. 26.

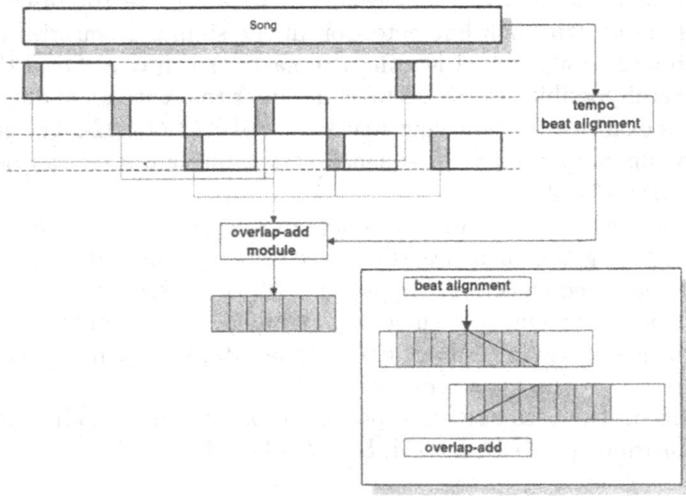

Fig. 26. Audio summary construction from sequence/state representation; details of fragments alignment and overlap-add based on tempo detection/ beat alignment.

6 Discussion

6.1 Comparing "Sequence" and "State" Approach

The "sequence" approach aims at detecting the repetition of sequences in the music, i.e. detecting two identical succession of events in the music such that each event of the first sequence is similar to its equivalent event in the second sequence but not to the other events in the sequence[8]. An example of sequence repetition in the case of popular music is the repetition of the successive notes of a melody.

The "state" approach aims at representing the music as a succession of states, such that a state is composed by a set of contiguous times with (somehow) similar events, and that two set of contiguous times of the music belonging to the same state represent (somehow) similar events.

In the state approach, all analysis times are attached to a state and it is possible that a state appears only once. In the sequence approach, only times belonging to a line detected in the similarity matrix belong to a sequence. A sequence appears at least twice (original and repetition). Because of this required lines detection step in the sequence approach, the sequence approach is less robust than the state approach. It is also computationally more expensive (need for a highest temporal resolution, need to compute a similarity matrix, need to perform the computationally expensive line detection process).

6.2 Evaluation

Evaluating the quality of the results obtained by the two approaches is a difficult task. The comparison of the structure obtained with the proposed algorithms with the "true" structure involves first deciding on what is the "true" structure. This last point is often subject of controversy among people (when is a melody repeated exactly, when is it a variation, when does this variation make it a different one ?). An example of this occurs in the title "Oh so quiet" from "Bjork". However the results obtained so far were found good by users.

Among the various types of audio summary, the most reliable one was found to be the "each" method. This is due to the fact that the longest/most-frequent sequence/state in a piece of music is not necessarily the most important music key for listeners. Moreover, deciding on a unique sequence/state makes the summary more sensitive to algorithm errors. The summary provided by the "all" method can be long and is redundant; however it is the only one that can remind to listeners the overall temporal structure of the piece of music. The "each" method, since it provides all important music key, rarely fails to provide the listener's preferred music key.

6.3 Limitations of Music Structure Discovery from Signal Analysis

Deriving content information directly from the signal implies a severe limitation: we do not observe the intention of the piece of music (the score) but (one

[8] Therefore a sequence representation cannot be derived from a state representation since all events inside a state are supposed to be similar.

of) its realization(s). Moreover, we only observe it through what we can actually automatically extract from it (considering the current limitations of digital signal processing). This implies especially limitations concerning multiple pitch estimation (chords estimation) and mixed sources (several instruments playing at the same time). Because of that, we will also be limited in deriving a single structure for the whole set of instruments (considering current source separation algorithm development, it is still difficult to separate the various instruments hence to obtain the structure for each instrument independently). Therefore, caution must be taken when concluding in the ability of a specific method to derive the actual musical structures. Indeed, except for music for which the musical arrangements (instruments playing in the background) are correlated with the melody part, or when the melody part is mixed in the foreground, there is little chance to be able to derive the actual melody repetitions. Moreover, since most music structure discovery methods are based on a search for repetitions, evolution of motives or of melodies are unlikely to be discovered.

7 Conclusion

In this paper we studied a "sequence" and "state" representation for music structure detection with the aim of generating visual and audio summaries. We introduced dynamic features, which seem to allow deriving powerful information from the signal for both 1) detection of sequences repetition in the music (lower/upper diagonals in a similarity matrix) and 2) representation of the music in terms of "states".

We proposed a 2D structuring filter algorithm for lines detection in the lag matrix and an algorithm to derive a sequence representation from these lines. We proposed a multi-pass algorithm based on segmentation and unsupervised learning (fuzzy-kmeans and HMM) for state representation of the music. We finally investigated sequential summaries generation from both sequential and state representations.

8 Perspectives

Combining both segment and state approach: Further work will concentrate on combining both sequence and state approaches (by using for example two different window length, it is possible to obtain both representation at the same time). It is clear that both approaches can help each other since the probability of observing a given sequence at a given time is not independent from the probability of observing a given state at the same time.

Hierarchical summaries: Further works will also concentrate on the development of hierarchical summary [16]. Depending on the type of information desired, the user should be able to select the "level" in a tree structure representing the piece of music (for example the various sub-melodies composing a melody or the various parts of a chorus/ verse).

Acknowledgments

Part of this work was conducted in the context of the European I.S.T. project CUIDADO [30] (http://www.cuidado.mu). Thanks to Amaury La Burthe (during its stay at Ircam), Kasper Souren and Xavier Rodet for the fruitful discussions .

References

1. J.-J. Aucouturier and M. Sandler. Segmentation of musical signals using hidden markov models. In *AES 110th Convention*, Amsterdam, The Netherlands, 2001.
2. J.-J. Aucouturier and M. Sandler. Finding repeating patterns in acoustic musical signals: applications for audio thumbnailing. In *AES 22nd International Conference on Virtual, Synthetic and Entertainment Audio*, Espoo, Finland, 2002.
3. M. Bartsch and G. Wakefield. To catch a chorus: Using chroma-based representations for audio thumbnailing. In *WASPAA*, New Paltz, New York, USA, 2001.
4. T. Beatles. Love me do (one, the best of album). Apple, Capitol Records, 2001.
5. Bjork. It's oh so quiet (post album). Mother records, 1995.
6. E. Cambouropoulos, M. Crochemore, C. Iliopoulos, L. Mouchard, and Y. Pinzon. Algorithms for computing approximate repetitions in musical sequences. In R. R. Simpson and J., editors, *10th Australasian Workshop On Combinatorial Algorithms*, pages 129–144, Perth, WA, Australia, 1999.
7. M. Cooper and J. Foote. Automatic music summarization via similarity analysis. In *ISMIR*, Paris, France, 2002.
8. T. Crawford, C. Iliopoulos, and R. Raman. String matching techniques for musical similarity and melodic recognition. In *Computing in Musicology*, volume 11, pages 73–100. MIT Press, 1998.
9. R. Dannenberg. Pattern discovery techniques for music audio. In *ISMIR*, Paris, 2002.
10. I. Deliege. A perceptual approach to contemporary musical forms. In N. Osborne, editor, *Music and the cognitive sciences*, volume 4, pages 213–230. Harwood Academic publishers, 1990.
11. J. Eckman, S. Kamphorts, and R. Ruelle. Recurrence plots of dynamical systems. *Europhys Lett*, (4):973–977, 1987.
12. J. Foote. Automatic audio segmentation using a measure of audio novelty. In *ICME (IEEE Int. Conf. Multimedia and Expo)*, page 452, New York City, NY, USA, 1999.
13. J. Foote. Visualizing music and audio using self-similarity. In *ACM Multimedia*, pages 77–84, Orlando, Florida, USA, 1999.
14. J. Foote. Arthur: Retrieving orchestral music by long-term structure. In *ISMIR*, Pymouth, Massachusetts, USA, 2000.
15. M. Hunt, M. Lennig, and P. Mermelstein. Experiments in syllable-based recognition of continuous speech. In *ICASSP*, pages 880–883, Denver, Colorado, USA, 1980.
16. A. Laburthe. *Resume sonore*. Master thesis, Universite Joseph Fourier, Grenoble, France, 2002.
17. K. Lemstrom and J. Tarhio. Searching monophonic patterns within polyphonic sources. In *RIAO*, pages 1261–1278, College of France, Paris, 2000.
18. B. Logan and S. Chu. Music summarization using key phrases. In *ICASSP*, Istanbul, Turkey, 2000.

19. Moby. Natural blues (play album). Labels, 2001.
20. MPEG-7. Information technology - multimedia content description interface - part 5: Multimedia description scheme, 2002.
21. Nirvana. Smells like teen spirit (nevermind album). Polygram, 1991.
22. N. Orio and D. Schwarz. Alignment of monophonic and polyphonic music to a score. In *ICMC*, La Habana, Cuba, 2001.
23. G. Peeters, A. Laburthe, and X. Rodet. Toward automatic music audio summary generation from signal analysis. In *ISMIR*, Paris, France, 2002.
24. L. Rabiner. A tutorial on hidden markov model and selected applications in speech. *Proccedings of the IEEE*, 77(2):257–285, 1989.
25. S. Rossignol. *Segmentation et indexation des signaux sonores musicaux*. Phd thesis, Universite Paris VI, Paris, France, 2000.
26. E. Scheirer. Tempo and beat analysis of acoustic musical signals. *JASA*, 103(1):588–601, 1998.
27. K. Souren. Extraction of structure of a musical piece starting from audio descriptors. Technical report, Ircam, 2003.
28. G. Tzanetakis and P. Cook. Multifeature audio segmentation for browsing and annotation. In *WASPAA*, New Paltz, New York, USA, 1999.
29. D. VanSteelant, B. DeBaets, H. DeMeyer, M. Leman, S.-P. Martens, L. Clarisse, and M. Lesaffre. Discovering structure and repetition in musical audio. In *Eurofuse*, Varanna, Italy, 2002.
30. H. Vinet, P. Herrera, and F. Pachet. The cuidado project. In *ISMIR*, Paris, France, 2002.
31. H. Zhang, A. Kankanhalli, and S. Smoliar. Automatic partitioning of full-motion video. *ACM Multimedia System*, 1(1):10–28, 1993.

Musical Style Classification from Symbolic Data: A Two-Styles Case Study

Pedro J. Ponce de León and José M. Iñesta

Departamento de Lenguajes y Sistemas Informáticos, Universidad de Alicante
Ap. 99, E-03080 Alicante, Spain
{pierre,inesta}@dlsi.ua.es

Abstract. In this paper the classification of monophonic melodies from two different musical styles (Jazz and classical) is studied using different classification methods: Bayesian classifier, a k-NN classifier, and self-organising maps (SOM). From MIDI files, the monophonic melody track is extracted and cut into fragments of equal length. From these sequences, A number of melodic, harmonic, and rhythmic numerical descriptors are computed and analysed in terms of separability in two music classes, obtaining several reduced descriptor sets. Finally, the classification results for each type of classifier for the different descriptor models are compared. This scheme has a number of applications like indexing and selecting musical databases or the evaluation of style-specific automatic composition systems.

Keywords: music information retrieval, self-organising maps, bayesian classifier, nearest neighbours (k-NN), feature selection.

1 Introduction

The automatic machine learning and pattern recognition techniques, successfully employed in other fields, can be also applied in music analysis. One of the tasks that can be posed is the modelization of the music style. Immediate applications are the classification, indexation and content-based search in digital music libraries, where digitised (MP3), sequenced (MIDI) or structurally represented (XML) music can be found. The computer could be trained in the user musical taste in order to look for that kind of music over large musical databases. Such a model could also be used in cooperation with automatic composition algorithms to guide this process according to a stylistic profile provided by the user.

Our aim is to develop a system able to distinguish musical styles from a symbolic representation of a melody using musicological features: melodic, harmonic and rhytmic ones. Our working hypothesis is that melodies from a same musical genre may share some common features that permits to assign a musical style to them. For testing our approach, we have initially chosen two music styles, jazz and classical, for our experiments. We will also investigate whether such a representation by itself has enough information to achieve this goal or, on the contrary, also timbric information has to be included for that purpose.

U.K. Wiil (Ed.): CMMR 2003, LNCS 2771, pp. 167–178, 2004.

In this work we will start with some related works in the area of musical style identification. Then our methodology will be presented, describing the musical data and the statistical description model we have used. Next, the implementation and parametrization for the different classification methods we used, namely Bayesian classifier, k-nearest neigbours (kNN) and self-organising maps (SOM) will be briefly explained. The initial set of descriptors will be statistically analized to test their contribut ion to the musical style separability. These procedures will lead us to reduced models, discarding not useful descriptors. Then, the SOM training and the classification results obtained with each classifier will be presented. These results are compared and the advantages and drawbacks of these classification methods related to the musical classification task are discussed. Finally, conclusions and current and future lines of work are presented.

1.1 Related Work

A number of recent papers explore the capabilities of SOM to analyse and classify music data. Rauber and Frühwirth [1] pose the problem of organising music digital libraries according to sound features of musical themes, in such a way that similar themes are clustered, performing a content-based classification of the sounds. Whitman and Flake [2] present a system based on neural nets and support vector machines, able to classify an audio fragment into a given list of sources or artists. Also in [3], the authors describe a neural system to recognise music types from sound inputs. In [4] the authors present a hierarchical SOM able to analyse time series of musical events and then discriminate those events in a different musical context. In the work by Thom [5] pitch histograms (measured in semitones relative to the tonal pitch and independent of the octave) are used to describe blues fragments of the saxophonist Charlie Parker. The pitch frequencies are used to train a SOM. Also pitch histograms and SOM are used in [6] for musicological analysis of folk songs.

These works pose the problem of music analysis and recognition using either digital sound files or symbolic representations as input. The approach we propose here is to use the symbolic representation of music that will be analysed to provide melodic, harmonic and rhythmic descriptors as input to the Bayesian, kNN and SOM classifiers (see Fig. 1) for classification of musical fragments into one of two musical styles. We use standard MIDI files as the source of monophonic melodies.

2 Methodology

2.1 Musical Data

MIDI files from two styles, jazz and classical music, were collected. Classical music was chosen and melodic samples were taken from works by Mozart, Bach, Schubert, Chopin, Grieg, Vivaldi, Schumann, Brahms, Beethoven, Dvorak, Haendel, Pagannini and Mendhelson. Jazz music consist of jazz standard tunes from a variety of authors like Charlie Parker, Duke Ellington, Bill Evans, Miles Davis, etc. The monophonic melodies are isolated from the rest of the musical content

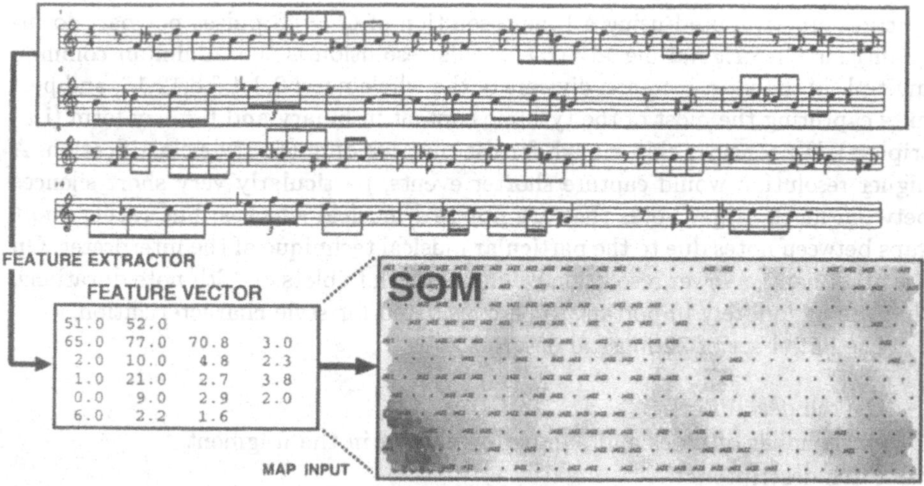

Fig. 1. Structure of the system: musical descriptors are computed from a window 8-bar wide and provided to a classifier (a SOM in this figure) for training and classification. Once trained, a style label is assigned to the units. During classification, the label of the winning unit provides the style to which the music fragment belongs to. This example is based on the Charlie Parker's jazz piece "Dexterity".

in these MIDI files —the track number of the melody been known for each file—. This way, we get a sequence of musical events that can be either notes or silences. Other kind of MIDI events are filtered out.

Here we will deal only with melodies written in 4/4. In order to have more restricted data, each melody sequence is divided into fragments of 8 bars. We assume this length suffices to get a good sense of the melodic phrase in the context of a 4/4 signature. Each of these fragments becomes a data sample. No segmentation techniques were applied to obtain the samples. You can compare this straightforward process as tuning a radio station, hear the music for a while and then switching off your radio box. Most people would be able to identify the music style they heard (provided they are familiar with that music style).

Each melodic sample is represented by a number of statistical descriptors (see section 2.2). The total number of samples is 1244, and approximately a 40% are jazz samples and a 60% are classical samples.

2.2 Feature Extraction

We have chosen a vector of musical descriptors of the melodic fragments as the input for the classifiers, instead of the explicit representation of the melodies. Thus, a description model is needed. Firstly, three groups of features are extracted: melodic, harmonic and rhythmic properties.

The melody tracks in the MIDI files are quantised in origin at 480 pulses per bar (i.e., real-time quantisation), therefore possibly containing some amount of phrasing like swing in jazz tunes, or stacatto/legato parts in classical music. The features are computed using a time resolution of $Q = 48$ pulses per bar, down-quantising the original melody track. This resolution is the minimum common multiple of the most common divisors of the whole note: 2,3,4,6,8,12,16, and permits capturing the most of the rythmic content in binary and ternary form (i.e., triplets). We consider this enough for the purpose of music style classification. A higher resolution would capture shorter events, particularly very short silences between much larger notes that are not real musical silences, but rather short gaps between notes due to the particular musical technique of the interpreter. On the other hand, a lower resolution would miss 8th triplets or 16th note durations, that are undoubtely important to be considered for style characterization.

The initial set of 22 musical descriptors is:

- Overall descriptors:
  - Number of notes and number of silences in the fragment.
- Pitch descriptors:
  - Lowest, highest (provide information about the pitch range of the melody), average, and standard deviation (provide information about how the notes are distributed in the score).
- Note duration descriptors (these descriptors are measured in pulses):
  - Minimum, maximum, average, and standard deviation.
- Silence duration descriptors (in pulses):
  - Minimum, maximum, average, and standard deviation.
- Interval descriptors (distance in pitch between two consecutive notes):
  - Minimum, maximum, average, and standard deviation.
- Harmonic descriptors:
  - *Number of non diatonic notes.* An indication of frequent excursions outside the song key (extracted from the MIDI file) or modulations.
  - *Average degree of non diatonic notes.* Describes the kind of excursions. Its a number between 0 and 4 that indexes the non diatonic notes of the diatonic scale of the tune key, that can be major or minor key[1]. It can take a fractional value.
  - *Standard deviation of degrees of non diatonic notes.* Indicates a higher variety in the non diatonic notes.
- Rhytmic descriptor: *number of syncopations*: notes not beginning at the rhythm beats but in some places between them (usually in the middle) and that extend across beats. This is actually an estimation of the number of syncopations. We are not interested in the exact number of them for our task.

In section 2.4 a feature selection procedure is presented in order to obtain some reduced description models and to test their classification ability.

We used the key meta-event present in each MIDI file to compute the harmonic descriptors. The correctness of its value was verified for each file prior to the feature extraction process.

[1] 0: ♭II, 1: ♭III (♮III for minor key), 2: ♭V, 3: ♭VI, 4: ♭VII.

With this set of descriptors, we assume the following hypothesis: melodies of the same style are closer to each other in the description space than melodies from different styles. We will test the performance of different classifiers to verify this hypothesis.

This kind of statistical description of musical content is sometimes referred to as *shallow structure description* [7]. It is similar to histogram-based descriptions, like the one found in [6], that it tries to model the distribution of musical events in a musical fragment. Computing the minimum, maximum, mean and standard deviation from the distribution of musical features like pitches, durations, intervals and non-diatonic notes we reduce the number of features needed (each histogram may be made up of tens of features), while assuming the distributions for the mentioned features to be normal within a melodic fragment. Other authors have also used some of the descriptors presented here to classify music [8].

2.3 Classifier Implementation and Tunning

Three different classifiers are used in this paper to automatic style identification. Two of them are supervised methods: The bayesian classifier and the kNN classifier [9]. The other one is an unsupervised learning neural network, the self-organising map (SOM) [10]. SOM are neural methods able to obtain approximate projections of high-dimensional data distributions into low-dimensional spaces, usually bidimensional. With the map, different clusters in the input data can be located. These clusters can be semantically labelled to characterise the training data and also hopefully future new inputs.

For the Bayesian classifier, we assume that individual descriptor probability distributions for each style are normal, with means and variances estimated from the training data. This classifier computes the squared Mahalanobis distance from test samples to the mean vector of each style in order to obtain a classification criterion. The kNN classifier uses an euclidean metrics to compute distance between samples, and we tested a number of odd values for k, ranging from 1 to 25.

For SOM implementation and graphic representations the SOM_PAK software [11] has been used. For the experiments, two different geometries were tested: 16×8 and 30×12 maps. An hexagonal topology for unit connections and a bubble neighbourhood for training have been selected. The value for this neighbourhood is equal for all the units in it and decreases as a function of time. The training was done in two phases. See Table 1 for the different training parameters. The metrics used to compute distance between samples was again the euclidean distance.

In the next pages, the maps are presented using the U-map representation, where the units are displayed by hexagons with a dot or label in their centre. The grey level of unlabelled hexagons represents the distance between neighbour units (the clearer the closer they are). For the labelled units is an average of the neighbour distances. This way, clear zones are clusters of units, sharing similar weight vectors. The labels are a result of callibrating the map with a series of

Table 1. SOM training parameters.

Map size	16 × 8	30 × 12
First training phase (coarse ordering)		
Iterations	3,000	10,000
Learning rate	0.1	0.1
Neigbourhood radius	12	20
Second training phase (fine tunning)		
Iterations	30,000	100,000
Learning rate	0.05	0.05
Neigbourhood radius	4	6

test samples and indicate the class of samples that activates more times each unit.

2.4 Feature Selection Procedure

We have devised a selection procedure in order to keep those descriptors that actually contribute to make the classification. The procedure is based on the values for all the features in the weight vectors of five previously trained SOMs. The maps are trained and labelled (calibrated) in an unsupervised manner (see Fig 2-a for an example). We try to find which descriptors provide more useful information for the classification. Some descriptor values for the weight vectors correlate better than others with the label distribution in the map. It is reasonable to consider that these descriptors contribute more to achieve a good separation between classes. See Fig. 2-b and 2-c for descriptor planes that correlate and that do not with the class labels.

Consider that the N descriptors are random variables $\{x_i\}_{i=1}^{N}$ that corresponds to the weight vector components for each of the M units in the map. We drop the subindex i for clarity, because all the discussion is related to each descriptor. We will divide the set of M values for each descriptor into two subsets: $\{x_j^C\}_{j=1}^{M_C}$ are the descriptor values for the units labelled with the classical style and $\{x_j^J\}_{j=1}^{M_J}$ are those for the jazz units, being M_C and M_J the number of units labelled with classical and jazz labels, respectively. We want to know whether these two set of values follow the same distribution or not. If false, it is an indication that there is a clear separation between the values of this descriptor for the two classes, so it is a good feature for classification and should be kept in the model and otherwise it does not seem to provide separability to the classes.

We assumed that both sets of values hold normality conditions and the following statistical for sample separation has been applied:

$$z = \frac{|\bar{x}_C - \bar{x}_J|}{\sqrt{\frac{s_C^2}{M_C} + \frac{s_J^2}{M_J}}} \quad , \tag{1}$$

where $\bar{x}_C$ and $\bar{x}_J$ are the means, and s_C^2 and s_J^2 the variances for the descriptor values for both classes. The larger the z value is, the higher the separation between both sets of values is for that descriptor. This value permits to order the

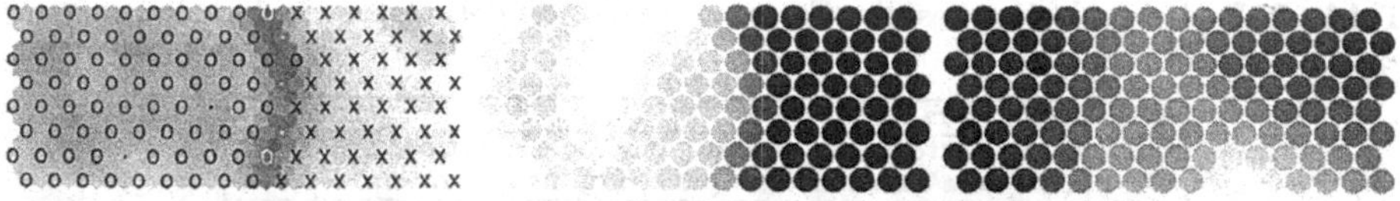

Fig. 2. Contribution to classification: (a:left) callibrated map ('X' and 'O' are the labels for both styles); (b:center) weight space plane for a feature that correlates with the areas; (c:right) plane for a feature that does not correlate.

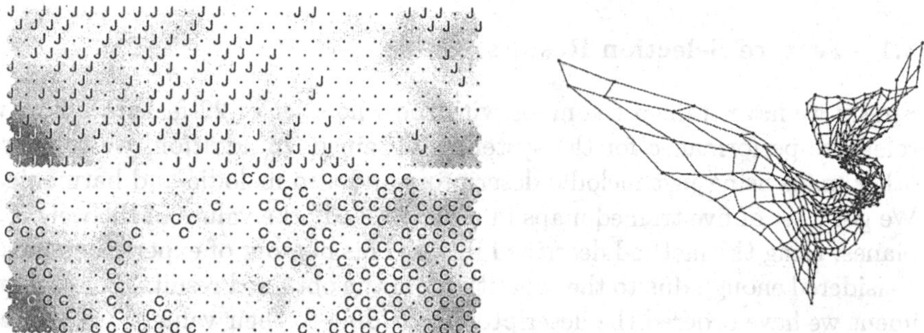

Fig. 3. Left: SOM map after being labeled with jazz (top) and classical (down) melodies. Note how both classes are clearly separated. Right: Sammon projection of the SOM, a way to display in 3D the organisation of the weight vector space.

descriptors according to their separation ability and a threshold can be established to determine which descriptors are suitable for the model. This threshold, computed from a t-student distribution with infinite degrees of freedom and a 99.5% confidence interval, is $z = 2.81$.

3 Experiments and Results

As stated above, we have chosen two given music styles: jazz and classical for testing our approach. The jazz samples were taken from jazz standards from different jazz styles like be-bop, hard-bop, big-band swing, etc., and the melodies were sequenced in real time. Classical tunes were collected from a number of styles like baroque, romantic, renaissance, impressionism, etc.

The first step was to train SOMs with the whole set of descriptors. After training and labelling, maps like that in figure 3 have been obtained. It is observed how the labelling process has located the jazz labels mainly on the left zone, and those corresponding to classical melodies on the right. Some units can be labelled for both music styles if they are activated by fragments from both styles. In these cases there is always a winner label (the one displayed) according to the number of activations. The proportion of units with both labels is the overlapping degree, that for the presented map was very low (8.0 %), allowing a clear distinction between styles.

In the Sammon projection of the map in figure 3 a knot separates two zones in the map. The zone at the left of the knot has a majority presence of units labelled with the jazz label and the zone at the right is mainly classical.

Table 2. Feature selection results. For each model the descriptors included are shown in the right column.

Model	Descriptors
6	Highest pitch, max. interval, dev. note duration, max. note duration, dev. pitch, avg. note duration
7	+syncopation
10	+avg. pitch, dev. interval, number of notes
13	+number of silences, min. interval, num. non-diatonic notes
22	All the descriptors computed

3.1 Feature Selection Results

Firstly we have trained the maps with the whole set of 22 features. This way a reference performance for the system is obtained. In addition, we have trained other maps using just melodic descriptors and also melodic and harmonic ones. We get a set of five trained maps in order to study the values of the weight space planes, using the method described in 2.4. This number of experiments has been considered enough due to the repetitivity of the obtained results. For each experiment we have ordered the descriptors according to their value for z_i (see eq. 1). In table 2 the feature selection results are displayed, showing what descriptors have been considered for each model according to those results. Each model number denotes the number of descriptors included in that model. We have chosen four reduced model sizes: 6, 7, 10 and 13 descriptors. The 7-descriptor model includes the best rated descriptors. The 6-descriptor model excludes syncopation. The other two models include other worse rated descriptors.

3.2 Classification

For obtaining reliable results a scheme based on *leave-k-out* has been carried out. In our case $k = 10\%$ of the size of the whole database. This way, 10 sub-experiments were performed for each experiment and the results have been averaged. In each experiment the training set was made of a different 90% of the total database and the other 10% was kept for testing.

The results obtained with the Bayesian classifier are presented in table 3. The best success rate was obtained with the 13-descriptor model (85.7%). Results from kNN classification can be seen in figure 4, with the error rate in function of the k parameter for each descriptor model. The minimum error rate was 0.113 (88.7% success) with $k = 9$ and a 7-descriptor model.

Table 3. Bayesian average success percentages.

Model	Jazz	Classic	Total
6	74.7	83.9	80.5
7	83.9	85.5	84.9
10	90.8	80.7	84.7
13	91.4	81.9	85.7
22	90.8	72.7	79.9

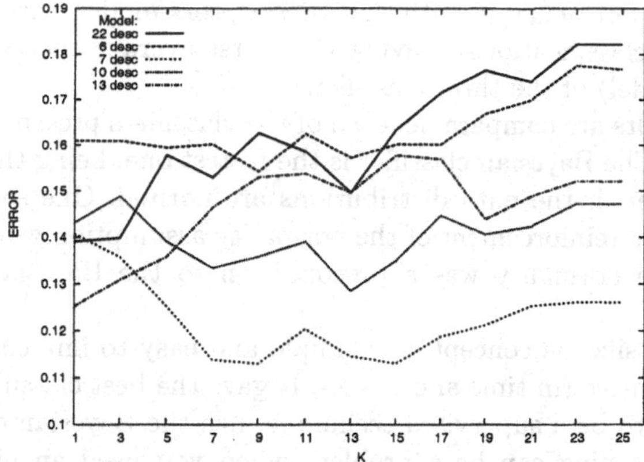

Fig. 4. kNN error rate with different k for each model.

Table 4. Average success rate percentages for the SOM classifier.

Model	Jazz	Classic	Total
6	79.4	82.1	80.8
7	81.8	86.5	84.2
10	78.8	82.7	80.7
13	72.0	82.6	77.3
22	72.7	79.6	76.1

Finally the results for the SOM classifier are presented in Table 4. These are average successful classification rate for Jazz and Classical samples. Each model has been evaluated with the two different SOM size SOM, and the best results presented here were consistently obtained with the 16×8 map geometry.

The best average results were obtained for that map when using the 7-descriptor model (84.2 %). It is observed that 6-descriptor model performance are systematically improved when syncopation is included in the 7-descriptor model. In some experiments even a 98.0 % of success (96.0 % for both styles) has been achieved. The inclusion of descriptors discarded in the feature selection test worsens the results and the worst case is when all of them are used (76.1 %). This is also true for the other classifiers presented here.

3.3 Result Comparison

The three classifiers give comparable results. The better average results were achieved with the kNN classifier with $k = 9$ and the 7-descriptors model (88.7% of average success). The kNN classifier provided an almost perfect classification in some of the leaving-10%-out partitions of the 7-descriptor model data, with $k = 7$ (99.1% of success, that means that only one melody out 123 was misclassified). The bayesian classifier performed similarly with some data partitions for 10 and 13-descriptors models reaching 98.4% of success (only two misclassifications), having ranked second in the average success rate (85.7%). The best results for SOM were achieved with the 7-descriptors model, with up to a 95.9% success (five misclassifications), having the worst average success rate (84.2%, 7-descriptors model) of the three classifiers.

All these results are comparable, each of the classifiers presenting advantages and drawbacks. The Bayesian classifier is the fastest one, being theoretically the optimum classifier, if the data distributions are normal. One secondary result we obtained is the reinforcement of the normality assumption we made to select descriptors, since normality was a precondition to the Bayesian classifier we implemented.

The *k*NN classifier is conceptually simple and easy to implement, but computationally intensive (in time and space). It gave the best classification results, while it remains to be a supervised technique, like the Bayesian classifier.

Supervised learning can be a problem when you need an high amount of training data to be reasonably sure that you are covering the most of the input space. This is our situation, where the input space, monophonic (or even polyphonic) melodies of arbitrary length from many different musical styles distribute in a very large input space. So we have focused our attention on the SOM, that provides results slightly worse than the other classifiers, but it has the advantage of being an unsupervised classification method (you need only a very few labelled samples to calibrate the map). Furthermore, SOMs provide additional information thanks to their data visualization capabilities.

As an example of the visualization capabilities of the SOM, Figures 5 and 6 show how an entire melody is located in the map. One melody of each style is shown, the first 64 bars of the Allegro movement from the *Divertimento in D* by Mozart for the classical style, and three choruses of the standard jazz tune *Yesterdays*. The grey area in each map corresponds to the style the melody belongs to. The first map picture for each melody shows the map unit activated by the first fragment of the melody. The next pictures show the map unit activated by the next fragment of that melody. Consecutively activated units are linked by a straight line, displaying the path followed by the melody in the map.

4 Conclusions and Future Works

We have shown the ability of three different classifiers to map symbolic representations of melodies into a set of musical styles using melodic, harmonic and rhythmic descriptions. The best recognition rate has been found with a 7-descriptor model where syncopation, note duration, and pitch have an important role.

Some of the misclassifications can be caused by the lack of a smart method for melody segmentation. The music samples have been arbitrarily restricted to 8 bars, getting just fragments with no relation to musical motives. This fact can introduce artifacts in the descriptors leading to less quality mappings. The main goal was to test the feasibility of the approach, and average recognition rates up to 88.7% have been achieved, that is very encouraging keeping in mind these limitations and others like the lack of valuable information for this task like timbre.

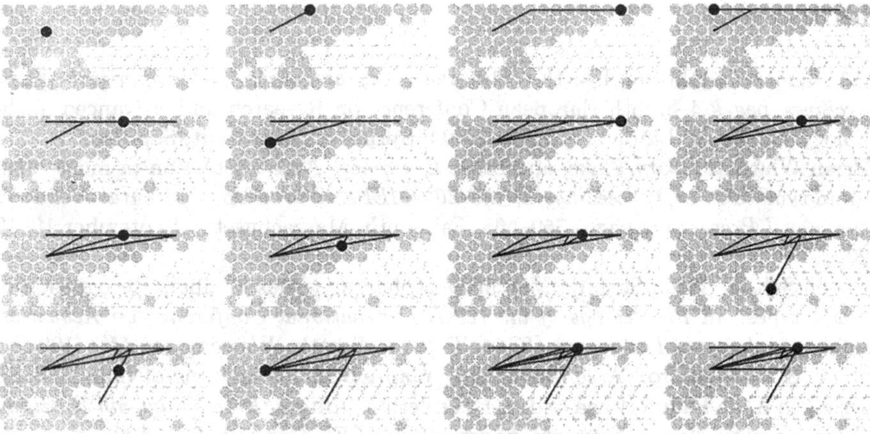

Fig. 5. Trajectory of the winning units for the *Divertimento in D (Mozart).*

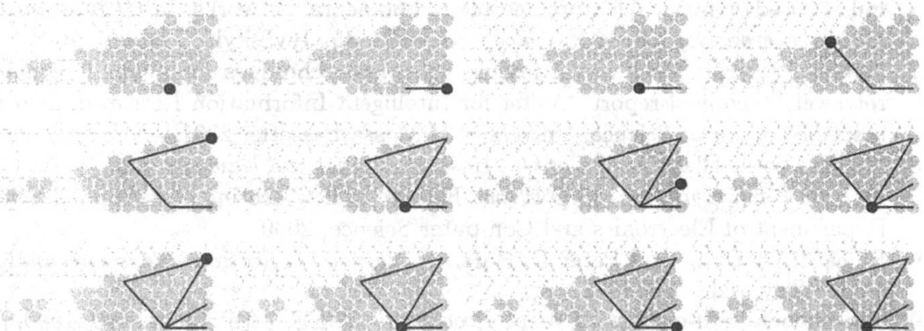

Fig. 6. Trajectory of the winning units for the jazz standard *Yesterdays.*

A number of possibilities are yet to be explored, like the development and study of new descriptors. It is very likely that the descriptor subset models are highly dependent on the styles to be discriminated. To achieve this goal a large music database has to be compiled and tested using our system for multiple different style recognition in order to draw significant conclusions.

The results suggest that the symbolic representation of music contains implicit information about style and encourage further research.

Acknowledgements

This paper has been funded by the Spanish CICyT project TAR, code: TIC2000–1703–CO3–02.

References

1. A. Rauber and M. Frühwirth. *Automatically analyzing and organizing music archives*, pages 4–8. 5th European Conference on Research and Advanced Technology for Digital Libraries (ECDL 2001). Springer, Darmstadt, Sep 2001.
2. Brian Whitman, Gary Flake, and Steve Lawrence. Artist detection in music with minnowmatch. In *Proceedings of the 2001 IEEE Workshop on Neural Networks for Signal Processing*, pages 559–568. Falmouth, Massachusetts, September 10–12 2001.
3. Hagen Soltau, Tanja Schultz, Martin Westphal, and Alex Waibel. Recognition of music types. In *Proceedings of the IEEE International Conference on Acoustics, Speech, and Signal Processing (ICASSP-1998)*. Seattle, Washington, May 1998.
4. O. A. S. Carpinteiro. A self-organizing map model for analysis of musical time series. In A. de Padua Braga and T. B. Ludermir, editors, *Proceedings 5th Brazilian Symposium on Neural Networks*, pages 140–5. IEEE Comput. Soc, 1998.
5. Belinda Thom. Unsupervised learning and interactive jazz/blues improvisation. In *Proceedings of the AAAI2000*, pages 652–657, 2000.
6. Petri Toiviainen and Tuomas Eerola. Method for comparative analysis of folk music based on musical feature extraction and neural networks. In *III International Conference on Cognitive Musicology*, pages 41–45, Jyväskylä, Finland, 2001.
7. Jeremy Pickens. A survey of feature selection techniques for music information retrieval. Technical report, Center for Intelligent Information Retrieval, Departament of Computer Science, University of Massachussetts, 2001.
8. Steven George Blackburn. *Content Based Retrieval and Navigation of Music Using Melodic Pitch Contours*. PhD thesis, Faculty of Engineering and Applied Science Department of Electronics and Computer Science, 2000.
9. Richard. O. Duda and Peter E. Hart. *Pattern classification and scene analysis*. John Wiley and Sons, 1973.
10. T. Kohonen. Self-organizing map. *Proceedings IEEE*, 78(9):1464–1480, 1990.
11. T. Kohonen, J. Hynninen, J. Kangas, and J. Laaksonen. Som_pak, the self-organizing map program package, v:3.1. Lab. of Computer and Information Science, Helsinki University of Technology, Finland, April, 1995. `http://www.cis.hut.fi/research/som_pak`.

FMF (Fast Melody Finder): A Web-Based Music Retrieval System

Seung-Min Rho and Een-Jun Hwang

The Graduate School of Information and Communication, Ajou University,
San 5, Wonchon-dong, Paldal-gu, Suwon 442-749, Korea
{anycall,ehwang}@ajou.ac.kr

Abstract. As the use of digital music is getting popular, there is an increasing demand for efficient retrieval of music. To do that, an effective music indexing and natural way of querying a music should be incorporated. This paper describes the FMF system[1] that designed to retrieve tunes from a database on the basis of a few notes which are drawn into a musical sheet applet or sung into a microphone. FMF system accepts both acoustic and visual input from the user, transcribes all the acoustic and common music notational inputs into specific strings such as UDR and LSR. Then, It searches an index for tunes that contain the sung pattern, or patterns similar to it. We implemented a web-based retrieval system and report on its performance through various experiments.

1 Introduction

Most traditional approaches for retrieving music are based on titles, composers or file names. But, due to their incompleteness or subjectiveness, it is hard to find music satisfying the particular requirements of applications. Even worse, such retrieval techniques cannot support queries such as "find music pieces similar to the one being played". Content-based music retrieval is usually based on a set of extracted audio features such as pitch, interval, duration and scale. One common approach for developing content-based music retrieval is to represent music as a string of characters using three possible values for the pitch change: U(p), D(own) and S(ame) or R(epeat). In order to search for the similar melodic string from melody sources, information retrieval techniques, especially string matching methods are used.

Some standard string matching algorithms such as Brute-Force, Knuth-Morris-Pratt or Boyer-Moore can find a certain sequence of strings in the text. Unfortunately, these algorithms find strings that match the input exactly. This is not suitable for acoustic input, since people do not sing or hum accurately, especially if they are inexperienced; even skilled musicians have difficulty in maintaining the correct pitch for the duration of a song. Therefore, it is common to use approximate matching algorithms instead of exact matching algorithms

[1] This work was supported by grant No. R05-2002-000-01224-0(2003) from the Basic Research Program of the Korea Science & Engineering Foundation.

U.K. Wiil (Ed.): CMMR 2003, LNCS 2771, pp. 179–192, 2004.

[6, 13, 17, 28, 29]. In general, approximate matching algorithms are far less efficient than exact matching algorithms. For that reason, we use both exact and approximate matching techniques. Approximate matching can be used where inaccuracy can be tolerated and exact matching should be considered where searching performance counts.

In this paper, we propose an enhanced audio retrieval system called *FMF* (Fast Melody Finder) using a set of frequently queried melody segments and various other audio features.

The rest of the paper is organized as follows. Section 2 presents an overview of the existing Music Information Retrieval (MIR) systems. In Section 3, we describe several music manipulation and comparison techniques used in the system. Section 4 presents our schema design for fast music retrieval using the index. Section 5 describes our prototype system and reports some of the experimental results and finally the last section concludes this paper.

2 Related Works

Introductory survey for indexing and retrieval of audio can be found in [6]. Ghias [5] developed a query by humming (QBH) system that is capable of processing an acoustic input in order to extract the necessary query information. However, this system used only three types of contour information to represent melodies. MELDEX (MELody inDEX) [17] system was designed to retrieve melodies from a database using a microphone. It first transforms acoustic query melodies into music notations, and then searches the database for tunes containing the hummed (or similar) pattern. This web-based system provides several match modes including approximate matching for interval, contour and rhythm. MelodyHound [27], originally known as "TuneServer," also used only three types of contour information to represent melodies. Tune recognition is based on error-resistant encoding and uses only the direction of the melody, ignoring the size of intervals or rhythm.

Foote [3] proposed a method for retrieving audio and music. Music retrieval is based on the classification of the music in the database. Classification of a piece of music is based on the distribution of the similarities. Blackburn [2] presented a new content and time based navigation tool. He extended the navigation to allow music-based navigation. Themefinder [10], created by David Huron, provides a web-based interface to the Humdrum thema command, which in turn allows database search for musical themes. It also allows user to find common themes in Western classical music and Folksongs of the 16th century.

3 Audio Feature Analysis and Representation

So far, most work has considered only UDR string parsed from pitch information to represent music. However, there are several restrictions in using the UDR string.

First, current UDR string cannot describe sudden pitch transitions. For example, as in Fig. 1, although the pitch contours for both bars are clearly different,

Fig. 1. Two different bars with the same pitch contour

they have the same string "UDU". Classifying intervals into five extended types could relieve this: up, up a lot, repeat, down and down a lot. The classification for up, down, and repeat is same as mentioned previously. The distinction between "down" and "down a lot" could depend on a certain threshold value on interval size, but a more reliable approach is to compare a note with a previously pitch contour. For instance, if the current note is lower pitch than the note before the last one, then it is classified as "down a lot." With this extension, music can now be parsed into a single string representation of the five pitch contour types: u, U, r, d and D for up, up a lot, repeat, down and down a lot, respectively. The first bar of the example can be represented as "udu" while the second bar as "UDU."

Second, pitch contour cannot represent information about note's length, tempo and scale; therefore, another representation for time contour should be considered. Similar to the way a pitch contour is described by the relative pitch transition, a time contour can be described by the duration between notes. Time in music is classified in one of three ways: R for a repetition of the previous time, L for longer than the previous time, or S for shorter than the previous time. For instance, in Fig. 2, the pitch contour and time contour of both bars are "udr" and "LSR". Nevertheless, they are actually different melody. It is because we did not consider the rhythm of the melody.

Fig. 2. Two different bars with the same time contour

Therefore, third, we have considered the rhythm of the melody using pitch name and duration of the note. Usual note pitch names - C, D, E, F, G, A, B - correspond to the musical scales "do re mi fa sol la si" and note durations - semiquaver, quaver, crotchet, minim and semibreve - are also represented by its corresponding weight "1", "2", "4", "8" and "16" respectively. In the case of dot note, add "." after the pitch name and add "#" and "-" in other cases like sharp and flat. For example, eight notes of the melody in Fig. 2 are coded by the following strings (assuming the basic length "1" is a semiquaver):

"F4 B8 G2 G2 F2 B4. G4 G4"

Suppose we compare this melody with a melody "F4 C4 G4 G4 F2 C4. G4 G4", which is stored in the database. However, we fail to identify the similarity between the two using an exact matching technique. To solve this problem, we use the Longest Common Subsequence (LCS) algorithm [1]. The LCS problem typically asks for solving on computing the length of some longest common subsequence. The LCS of the both melodies is "F4 G2 G2 F2 G4 G4".

4 *FAI* Indexing Scheme

This section describes a schema for describing music in the database and construction and maintenance of dynamic index for fast searching from frequently queried melody tunes.

4.1 *FAI* Indexing Scheme

Many file formats such as SMDL, NIFF and MIDI are used to store musical scores. A graphical format like Notation Interchange File Format (NIFF) is not suitable for general interchange, which is one of the main reasons NIFF has not been adopted by many applications. Standard Music Description Language (SMDL) was designed as a representational architecture for music information; however, there is no commercial software supporting it because of its overwhelming complexity. MP3 and other digital audio encoding formats represent music recordings, not music notation. Except for very simple music, computers cannot automatically derive accurate music notation from the music recordings, despite many decades of researches. On the other hand, MIDI file contains more score information than other formats. Also, it can hold other text data such as titles, composers, track names, or other descriptions. Therefore, we will use the Musical Instrument Digital Interface (MIDI) format file.

Fig. 3 shows the schema diagram for musical features mostly derived from the MIDI format. In the figure, the *Music* element has two components: total number of music and scores of music. Each *MusicScore* element has its id, meta-information, attributes and part of the scores of the music. The *MetaInfo* element has information such as title, composer and filename. The *Attributes* element describes musical attributes of a score, such as the clef, key and time signature. *Key signature* is represented by the number of sharps or flats. The *Time* element represents *Beats* and *BeatType* elements that are the numerator and denominator of the time signature respectively. The *Part* element contains its id and the number of phrases.

4.2 Dynamic Indexing

In the content-based applications, indexing and similarity-searching techniques are two keys to fast and successful data retrieval. Currently, acoustic features are

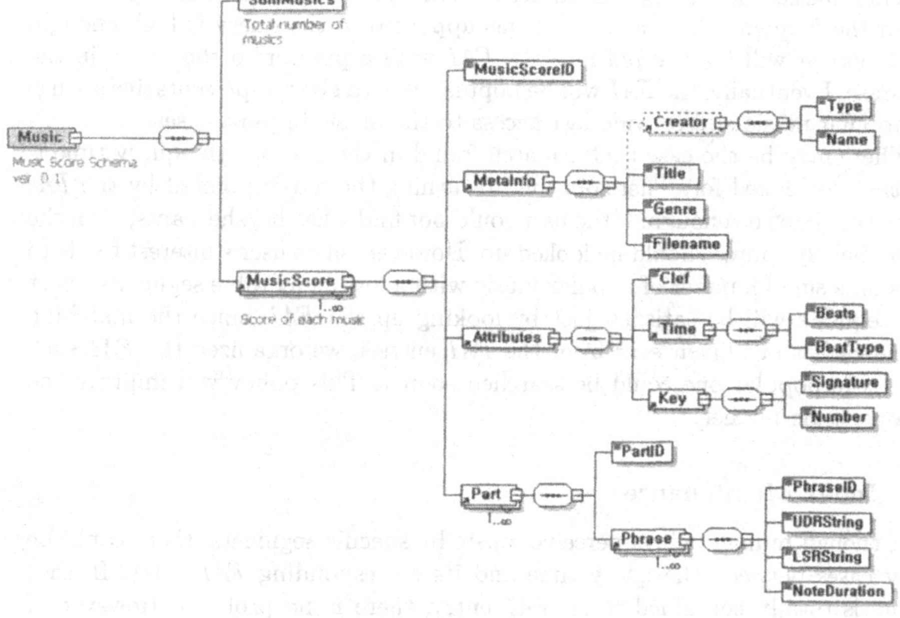

Fig. 3. XML schema for music meta-data

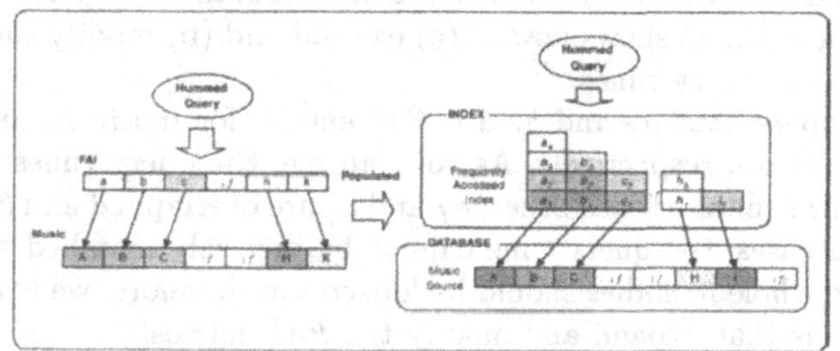

Fig. 4. *FAI* Indexing scheme

widely used for indices [3, 30]. In this paper, we use five types of UDR notation based on the pitch contour and LSR notation based on the time contour to represent music contents in the database.

Fig. 4 depicts the indexing scheme. As mentioned before, music is usually represented or memorized by a few specific melody segments. It means that most user queries will concentrate on those segments. If we use them as an index into music database, it would give great advantage in terms of response time. Those segments can be found from the previous user queries. We have organized them into an index structure called *FAI*. Initially, *FAI* could be empty or initialized with guessed segments for popular music. In this case, whole melody strings should be looked up for the query tune. Relying on the user response, matched music and its segment matched with the query tune will be recorded. When the frequency that a segment has appeared in the query is high enough, that segment will be inserted into the *FAI* with a pointer to the music in the database. Eventually, the *FAI* will be populated with short representative strings for popular music and provide fast access to the music in most cases.

There may be the case that a match found in the *FAI* for the query tune is not the one looked for. Therefore, just returning the music pointed by the *FAI* entry would not be enough. If the user could not find what he/she wants, then the entire melody source should be looked up. However, since user's interest tends to focus on a small number of popular music with a few memorable segments, most user requests will be satisfied just by looking up the *FAI*. Since the matching engine carries out linear search for the *FAI* entries, we organized the *FAI* such that more popular one could be searched sooner. This policy will improve the response time further.

4.3 Index Maintenance

Even though human beings perceive music by specific segments, there could be many cases between the query tune and its corresponding *FAI* entry. In case a tune is totally contained in an *FAI* entry, there is no problem. However, if they overlap, it could be considered a miss from the FAI search and cause the whole melody to be searched for. In order to avoid such unnecessary search in the future, FAI entries need to be expanded or modified properly according to user query tunes. Fig. 5 shows how to (a) expand and (b) modify the FAI entries depending on the query tunes.

Let us suppose that a_f and b_f are *FAI* entries for music A, and a_q and b_q are the query tunes, respectively. As you can see, the query tunes and the *FAI* entries are very similar to each other; a_f and a_q are overlapped and b_f is included in b_q. In both cases, the query tune cannot be precisely matched with the *FAI* entry, and the whole melodies should be looked up; therefore, we need to include some operations that expand and modify the *FAI* entries.

Fig. 5-(a) shows how to expand *FAI* entries by merging them with given query tunes. In Fig. 5-(b), however, both the query a_q and b_q are not a substring but a subsequence of *FAI* entries a_f and b_f. Therefore, we also need more operations such as insertion, deletion and substitution in addition to the operation of expansion. To compute the subsequence of the query tune and *FAI* entry, unnecessary characters should be deleted first and then both strings $a_f(b_f)$ and $a_q(b_q)$ are aligned from the first matched character. Strings $a_f(b_f)$ and $a_q(b_q)$ after the deletion and alignment are shown below.

a_f: D R D D R D D b_f: D U R R R U U
a_q: D R R D R D D b_q: D U R R R D U U

In this alignment, R in a_q is mismatched with D in a_f, and all other characters match to each other; therefore, the operation of substitution is required. Pseudo code for processing the query using both exact and approximate matching functions is shown in the appendix.

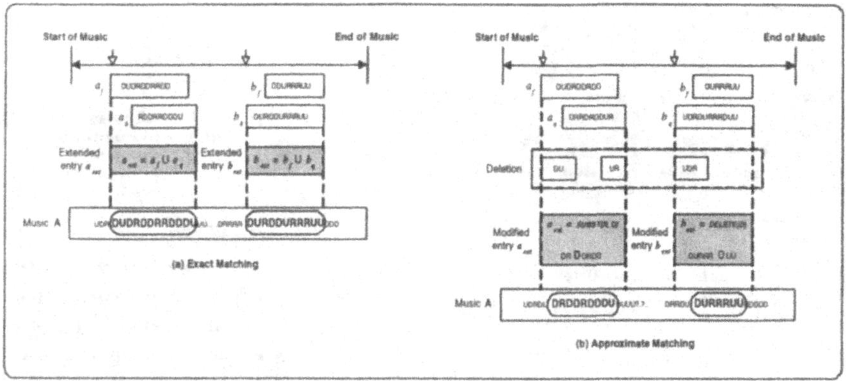

Fig. 5. Expanding and Modifying of the *FAI* entries

5 Implementation

This section describes the overall system architecture of our prototype system and briefly shows the interfaces for querying and browsing audio data, and evaluates its performance.

5.1 System Architecture

We have implemented a prototype music retrieval system based on the indexing scheme. It provides flexible user interface for query formulation and result browsing. Both client and server side are implemented by Java Applet, JSP. We used a set of jMusic[11] APIs for extracting audio features and the eXcelon database system for handling meta-data in XML. We have used approximately 12,000 MIDI files for the experiment and the average file size was about 40Kbytes. Fig. 6 depicts the overall system structure and the querying process.

Given a query either by humming with microphone or CMN-based interface, the system interprets audio input as a signal or a sequence of notes respectively and extracts features such as pitch and time contour. Then, extracted audio features are transcribed into UDR, LSR and pitch duration notation. After the transcription, it searches the *FAI* first before searching into the music database. If a match is found, it increments the frequency of the entry and returns it to the result set; otherwise, matching engine searches for entire melodies in database. If a similar melody is found from the database and the user confirms it, then the query tune is inserted into the *FAI* for that music.

5.2 Query and Browsing

Query interfaces for humming and CMN are shown in Fig. 7. The CMN-based query is created by mouse using drag and drop on the music sheet applet and humming-based query is produced by humming or singing on the microphone.

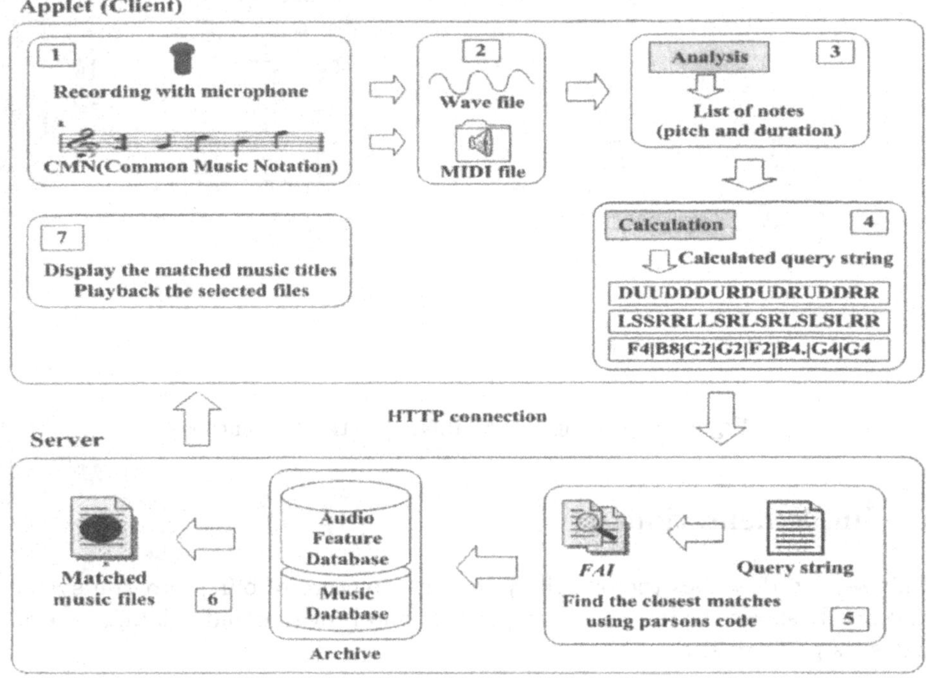

Fig. 6. System architecture and the querying process

Both queries are stored as a MIDI format and then transcribed into strings as described in Section 3.

In addition, the system allows query by composer, title, file name, and other text data with the same indexing scheme. All those text information are collected from the MIDI files as described. This enables keyword-based music search. That is, FMF can perform full-text search using the appropriate keywords and melody matching using the specified index for music.

Query melody depicts two queries based on the first theme occurrence in Christmas song "Silver Bells" and search interface is shown in Fig. 8.

Fig. 9 shows the number of matched files from the *FAI* and the music database from the queried melody in Fig. 8. User can play the MIDI file by clicking the hyperlinks.

5.3 Experiments

We used three exact matching algorithms and one approximate matching algorithm in our experiment. Three exact matching algorithms are Naive, Knuth-Morris-Pratt (KMP) and Boyer-Moore (BM). Boyer-Moore is the fastest on-line algorithm whose running time is O(n/m) where n is the length of the entire database. Because of its better performance, we have chosen Boyer-Moore al-

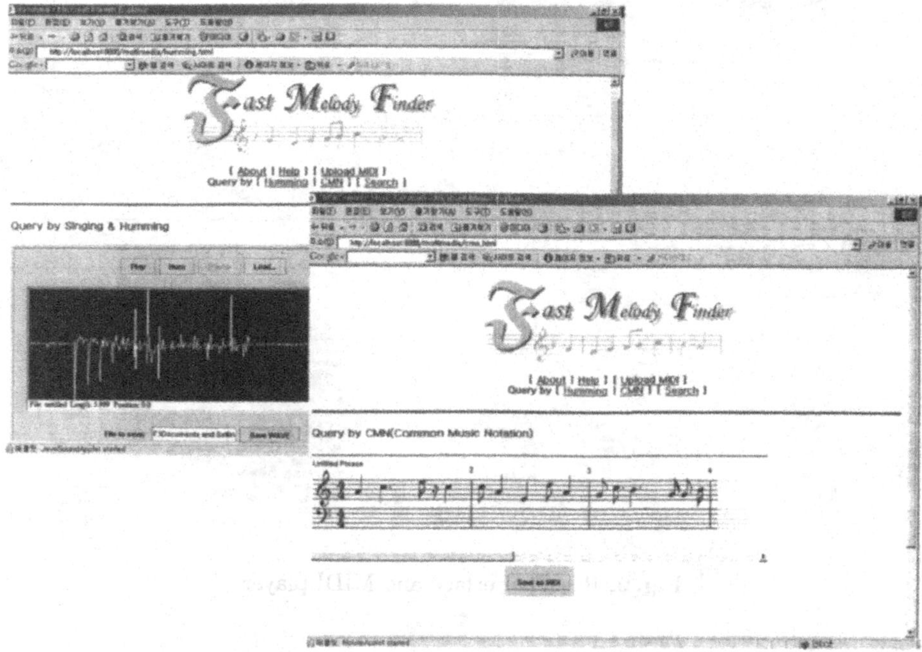

Fig. 7. Query interfaces of CMN and Humming

Fig. 8. Query melody and search interface

gorithm for exact matching in the system and edit distance with dynamic programming for approximate string matching.

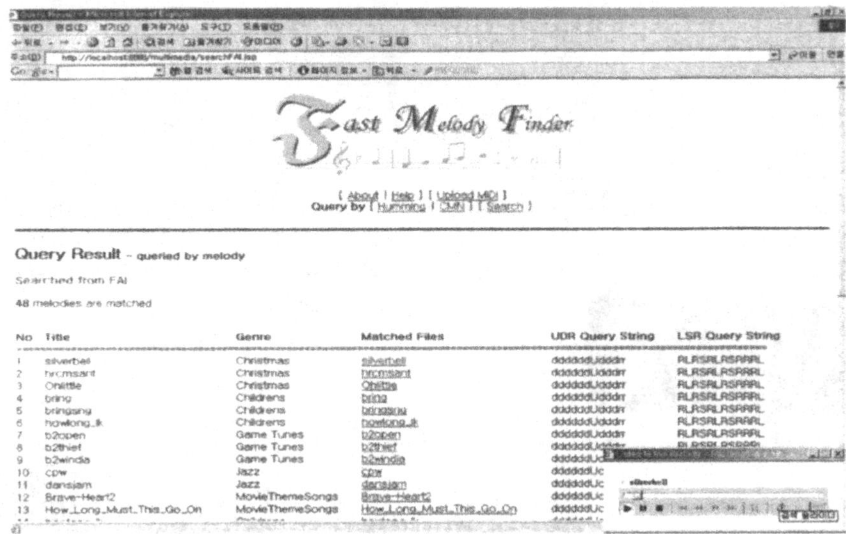

Fig. 9. Result interface and MIDI player

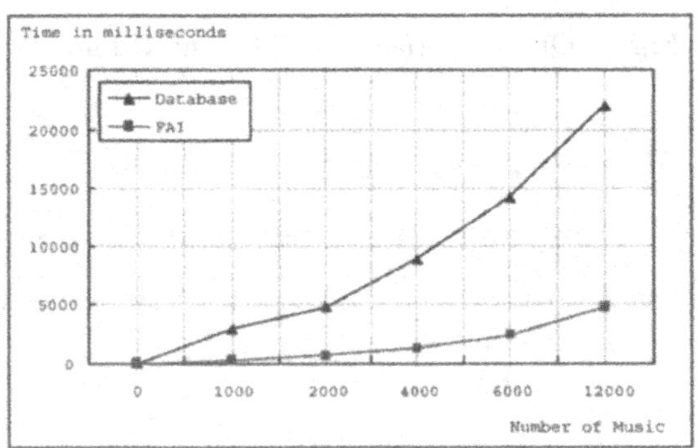

Fig. 10. Response time vs # of music in *FAI* and Database

Fig. 10 shows the average response time for user query. The upper curve represents the response time with usual index scheme and the lower curve represents the response time with *FAI*. In both cases, as the number of songs is increased, more time has been taken to find out matches from the database, but the *FAI* approach shows much better result even for a larger data set.

The first graph in Fig. 11 shows that query length does not seem to affect retrieval performance. The length of the query plays an important role in a larger and more varied collection. More matching files in fewer notes in query suggest that fewer patterns of notes in query are common in most music as shown in the second graph.

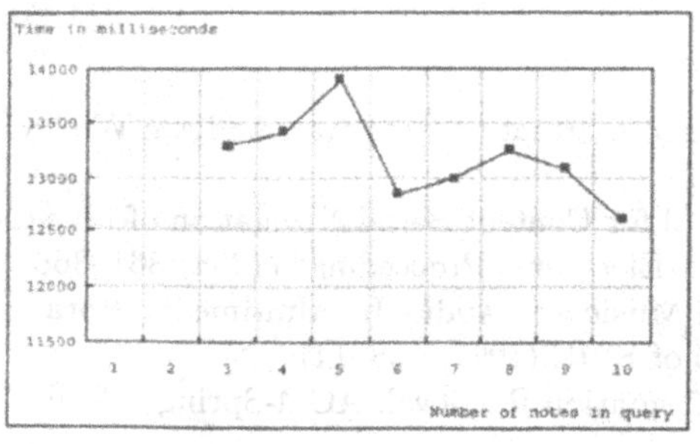

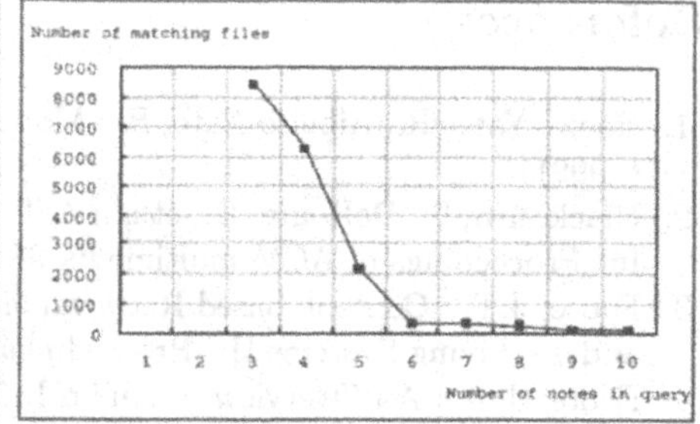

Fig. 11. Response time vs # of notes in query and # of matching melodies vs # of notes in query

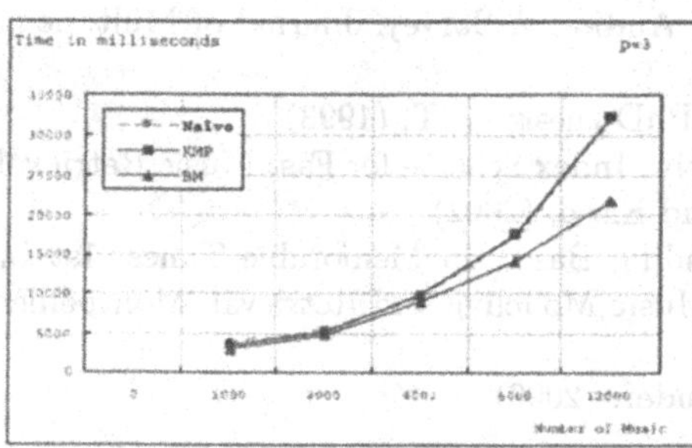

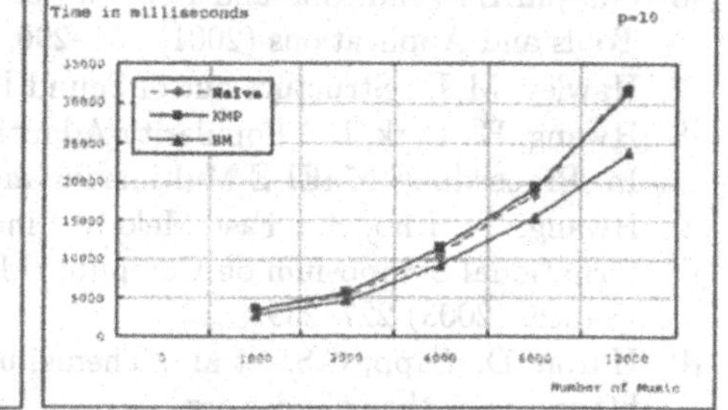

Fig. 12. Query response time for Naive, KMP and BM algorithm when p=3 and p=10

Fig. 12 shows the query response times for Naive, KMP and BM algorithms. The two upper curves represent the query response time for Naive and KMP. The lower curve represents the query response time for BM, which gives better performance than other two algorithms. We compared their performances using UDR strings with the length p=3 and p=10 by measuring the average response time.

6 Conclusion

In this paper, we discussed various features of music contents for content-based retrieval and proposed a fast music retrieval system based on frequently accessed tunes. A series of experiments performed on a prototype system show that our *FAI* technique gives much better performance to existing indexing approach. This is due to the fact user queries are usually concentrated to a small set of music and even query tunes for each music are focused on a few segments. For this reason, *FAI* takes only small amount of system resource such as memory and can be placed onto arbitrary music retrieval systems to improve the performance.

References

1. Baeza-Yates R., Ribeiro-Neto, B. : Modern Information Retrieval. Addison Wesley. (1999)
2. Blackburn, S., DeRoure, D., et al.: A Tool for Content Based Navigation of Music. In: Proceedings of ACM multimedia 98 - Electronic Proceedings (1998) 361–368.
3. Foote, J.T. : Content-Based Retrieval of Music and Audio. In: Multimedia Storage and Archiving Systems II - Proceedings of SPIE (1997) 138–147.
4. Foote, J.T. : An Overview of Audio Information Retrieval. ACM-Springer Multimedia Systems. (1998) 2–10.
5. Ghias, A., et al. : Query by humming - musical information retrieval in an audio database. In: Proceedings of ACM Multimedia 95 - Electronic Proceedings (1995) 231–236.
6. Guojun Lu : Indexing and Retrieval of Audio : A Survey. Journal of Multimedia Tools and Applications (2001) 269–290.
7. Hawley, M.J. : Structure out of Sound PhD thesis, MIT. (1993)
8. Hwang, E., Park, D. : Popularity-Adaptive Index Scheme for Fast Music Retrieval. In: Proceedings of IEEE Multimedia and Expo. (2002)
9. Hwang, E., Rho, S. : Fast Melody Finding Based on Memorable Tunes. 1st International Symposium on Computer Music Modeling and Retrieval, Montpellier, France. (2003) 227–239.
10. Huron, D., Sapp, C.S., et al. : Themefinder. (2000) http://www.themefinder.org
11. jMusic : Java library. http://jmusic.ci.qut.edu.au/
12. Kornstadt, A. : Themefinder: A Web-Based Melodic Search Tool. Computing in Musicology 11, MIT Press. (1998)
13. Kosugi, N., et al. : A Practical Query-By-Humming System for a Large Music Database. In: Proceedings of the 8th ACM International Conference. (2000) 333–342.
14. Lemstorm, K., et al. : Retrieving Music - To Index or not to Index. ACM International Multimedia Conference. (1998) 64–65.
15. Lemstorm, K., et al. : SEMEX - An efficient Music Retrieval Prototype. In: Proceeding of Symposium on Music Information Retrieval.
16. McNab, R.J., et al. : Toward the digital music library: tune retrieval from acoustic input. In: Proceedings of the first ACM Conference on Digital Libraries. (1996) 11–18.
17. McNab, R.J., et al. : The New Zealand digital library MELody index. Digital Libraries Magazine. (1997) 11–18.
18. Melucci, M., et al. : Musical Information Retrieval using Melodic Surface. In: Proceedings of the ACM Digital Libraries Conference. (1999) 152–160.
19. Michael, G. : Representing Music Using XML. International Symposium on Music Information Retrieval.
20. MiDiLiB project : Content-based indexing, retrieval, and compression of data in digital music libraries. http://www-mmdb.iai.uni-bonn.de/forschungprojekte/midilib/english/
21. Rolland, P.Y., et al. : Musical Content-Based Retrieval: an Overview of the Melodiscov Approach and System. In: Proceedings of ACM multimedia 98 - Electronic Proceedings. (1998) 81–84.

22. Rolland, P.Y., et al. : Music Information Retrieval: a brief Overview of Current and Forthcoming Research. In: Proceeding of Human Supervision and Control in Engineering and Music.
23. Salosaari, P., et al. : MUSIR-A Retrieval Model for Music. Technical Report RN-1998-1, University of Tampere. (1998)
24. Subramanya, S.R., et al. : Transforms - Based Indexing of Audio Data for Multimedia Databases. IEEE International Conference on Multimedia Systems. (1997)
25. Subramanya, S.R., et al. : Use of Transforms for Indexing in Audio Databases. International Conference on Computational Intelligence and Multimedia Applications. (1999)
26. Tseng, Y.H. : Content-Based Retrieval for Music Collections. In: Proceedings of the 22nd Annual International ACM SIGIR Conference on Research and Development in Information Retrieval. (1999) 176–182.
27. Typke, R., Prechelt, L. : An Interface for melody input. ACM Transactions on Computer-Human Interaction. (2001) 133–149.
28. Uitdenbogerd, A., Zobel, J. : Manipulation of music for melody matching. In: Proceedings of ACM Multimedia Conference. (1998) 235–240.
29. Uitdenbogerd, A., Zobel, J. : Melodic matching techniques for large music databases. In: Proceedings of ACM Multimedia Conference. (1999) 57–66.
30. Wold, E., et al. : Content-based Classification, Search and Retrieval of Audio. IEEE Multimedia 3(3) 27–36.

Appendix A

Query Process Algorithm in *FAI*

```
(1) Get a humming query from input device - microphone, CMN;
(2) While (any query comes into the system)
    If (Exact Matching is selected)
        Call ExactMatch();
    Else
        Call ApproxMatch();
(3) List up the candidate result melody ordered by the highest ranked melody
first;
(4) Play the retrieved melody;

/* Definition of Functions */
Function ExactMatch(query q)
    If (melody q is equal to the melody in FAI)
        Increase access_count value of FAI entry;
        Return matched_melody;
    Else
        ResultSet = search the whole music DB;
        If (same melody is found in ResultSet)
            Add a melody into FAI;
End Function
```

```
Function ApproxMatch(query q)
    Empty Matrix M for query q and string in FAI with i-th row and j-th col-
umn;
    M[0, j] = 0 and M[i, 0] = i;
    While (No errors found)
        If (melody q is equal to the melody in FAI)
            M[i, j] = M[i-1, j-1];
        Else
            M[i, j] = 1 + Min(M[i-1, j], M[i, j-1], M[i-1, j-1]);

    If (M[i, j] is less than Threshold)
        Increase access_count value of FAI entry;
        Return matched_melody;
    Else
        ResultSet = search the whole music DB;
        If (same melody is found in ResultSet)
            Add melody into FAI;
End Function
```

The Representation Levels of Music Information

Hugues Vinet

IRCAM
1, place Igor Stravinsky
F-75004 Paris, France
hugues.vinet@ircam.fr
http://www.ircam.fr

Abstract. The purpose of this article is to characterize the various kinds and specificities of music representations in technical systems. It shows that an appropriate division derived from existing applications relies in four main types, which are defined as the *physical, signal, symbolic* and *knowledge levels*. This fair simple and straightforward division provides a powerful grid for analyzing all kinds of musical applications, up to the ones resulting from the most recent research advances. Moreover, it is particularly adapted to exhibiting most current scientific issues in music technology as problems of *conversion between various representation levels*. The effectiveness of these concepts is then illustrated through an overview of existing applications functionalities, in particular from examples of recent research performed at IRCAM.

1 Introduction

The growing importance of the music industry as a key economic sector combined with the current convergence of computer, audiovisual and telecommunication technologies, yields rapid developments in music technologies. These technical evolutions have an impact at all levels of the production chain (production, publishing, dissemination and consumption), and bring new modalities of presentation, access to, and manipulation of the music material. As a result, there is an unprecedented variety of applications resulting from the music technology industry and research. Examples of such applications include score editors, MIDI and audio sequencers, real time DSP modules, virtual instruments based on physical modeling, computer-aided composition environments, 3D audio rendering systems, title databases with content-based browsing features, etc.

A question then arises: given the various approaches to the music phenomenon developed in these applications, is it possible to derive a global view, which integrates all kinds of associated representations in a single, unified scheme? In the context of this first issue of CMMR and its focus on music modeling issues, the purpose of this article is to answer this question by characterizing the specificities of music representations in technical systems. Therefore, the proposed approach relies on the identification of a limited number of well-defined representation types, called *Representation Levels*, or RLs, for reasons to be further developed, and to analyze existing applications through this RL grid. The effectiveness of these concepts will be then illustrated through an overview of existing applications features, in particular from recent research performed at IRCAM.

U.K. Wiil (Ed.): CMMR 2003, LNCS 2771, pp. 193–209, 2004.

2 Definitions and Properties of the Representation Levels

There are multiple ways of representing music information in technical systems, and in particular with computers. Such representations are chosen according to relevant viewpoints on the music content in order to match the target system functions. The term "Representation" refers here to the way information is represented internally in the system, i.e. essentially data structures. This article, in its aforementioned scope, does not handle another complementary aspect of representations in applications, related to man-machine interfaces, i.e. the way internal data are mediated to the user and, inversely, the way he can access them for manipulation. These issues are handled elsewhere in the context of man-machine interfaces for music production [23].

2.1 Music Representation Types

The conception of various kinds of music representation types is motivated by the recent history of music technology. There has been for several decades two main distinct and complementary ways of representing music content in technical systems:

- audio signal representations, resulting from the recording of sound sources or from direct electronic synthesis,
- symbolic representations, i.e. representations of discrete musical events such as notes, chords, rhythms, etc.

This distinction has been effective since the very beginning of computer music and is respectively exemplified by the pioneering works, almost simultaneous in the 1950s, of Max Matthews for the first digital music syntheses [10] and Lejaren Hiller for music compositional algorithms [7] . It is still true with current commercial music applications such as sequencers, in which digital audio and MIDI formats coexist.

Fundamental differences between both representations can be expressed as follows:

- the symbolic representation is *content-aware* and describes events in relation tofor-malized concepts of music (music theory), whereas the signal representation is a blind[1], content-unaware representation, thus adapted to transmit any, non-musical kind of sound, and even non-audible signals[2].
- even digitized through sampling, the signal representation appears as a continuous flow of information, both in time and amplitude, whereas the symbolic representation accounts for discrete events, both in time and in possible event states (e.g. pitch scales). Low bandwidth control parameters, such as MIDI continuous controllers, are also part of this category.

It should also be noted that despite various existing methods for coding audio signals, be them analog or digital, even in compressed form, they all refer to and enable the reconstruction of the same representation of audio signals as amplitude functions of time. On the contrary, symbolic representations gather a variety of descriptive approaches, including control-based information such as in the MIDI and General

[1] One should rather say *deaf* in this context, but languages provide few auditory metaphors...

[2] For instance, a sinusoidal function at a 1Hz frequency

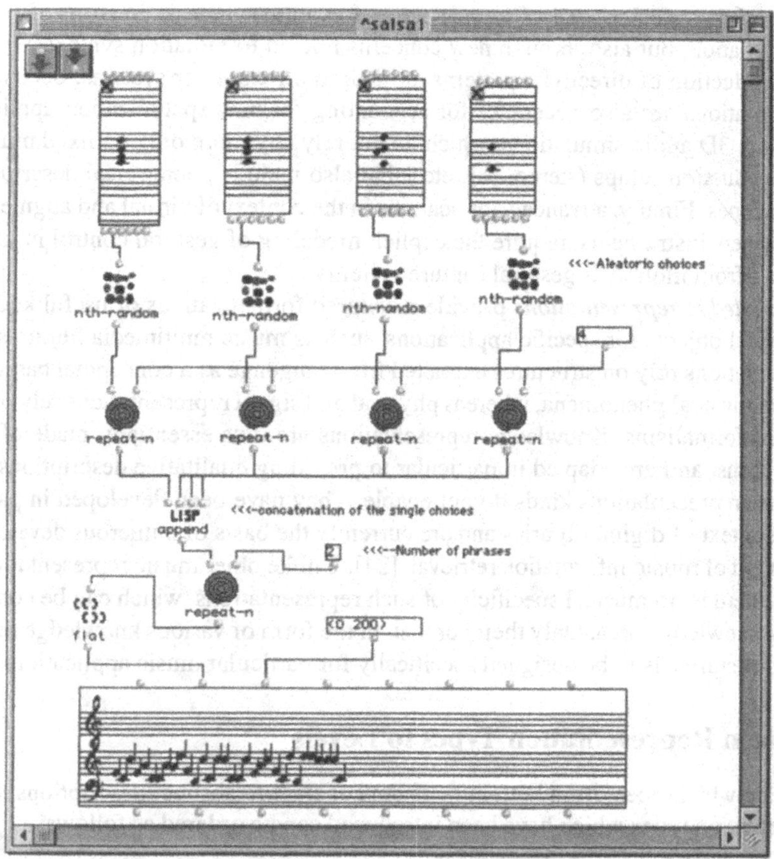

Fig. 1. Patch example in OpenMusic. The Input materials (chords) are positioned on top of the window, and the produced result is displayed in a notation editor at the bottom

MIDI standards, score descriptions used in score editors, or more sophisticated, object-oriented musical structures found in computer-aided composition environments. For instance, the OpenMusic environment, developed at IRCAM, is a visual programming environment which enables, as illustrated in Figure 1, to design processing functions of symbolic information [2].

However, these two kinds of representations are not sufficient for characterizing all aspects of music contents found in existing technology. In order to take into account recent advances in musical applications, it is necessary to integrate two other kinds of representations, hereinafter defined as *physical* and *knowledge* representations.

Physical representations result from physical descriptions of musical phenomena, in particular through acoustic models. As one of their specificities, these representations account for *spatial characteristics* of sound objects and scenes, in terms of geometrical descriptions, but also include other physical properties, e.g. mass, elasticity and viscosity. The introduction of physical representations is first motivated by the growing importance of physical models of sound sources for audio synthesis, in terms of excitation and oscillation, but also through new concerns related to radiation synthesis [13], i.e. the reproduction of directivity patterns through multi-excitator systems. Second, these representations are also necessary for accounting for new spatialization applications, including. 3D audio simulations, which do not rely any more only on fixed multichannel reproduction setups (stereo, 5.1, etc.), but also include geometrical descriptions of sound scenes. Finally, advanced applications in the context of virtual and augmented reality, or new instruments, require the explicit modeling of gestural control information resulting from motion or gestural capture systems.

Knowledge representations provide structured formalizations of useful knowledge on musical objects for specific applications, such as music multimedia libraries. These representations rely on structures extracted from language as a conceptual basis for describing musical phenomena, whereas physical and signal representations rely on mathematical formalisms. Knowledge representations are thus essentially made of textual descriptions, and are adapted in particular to providing qualitative descriptions, which the other representations kinds do not enable. They have been developed in particular in the context of digital libraries and are currently the basis of numerous developments in the field of music information retrieval [24]. Unlike other music representation categories, there is no musical specificity of such representations, which can be considered for any knowledge area. Only their content, in the form of various knowledge representation structures, is to be designed specifically for particular music applications.

3 From Representation Types to Levels

In a scale which goes, from bottom up, from concrete to abstract descriptions, the four representation types which have been introduced can be ordered as follows:

Table 1. Ordering of Music Representation Types

Knowledge Representations
Symbolic Representations
Signal Representations
Physical Representations

An analogy could be found between the resulting levels and the various stages of musical information processing by the brains, from the auditory system physiology up to the highest cognitive levels. In this analogy, which is valid up to a certain extent, the signal level would correspond to the binaural signals, i.e. the acoustic pressure signals at the level of both eardrums, which characterize information inputted into the auditory system. Subsequently, the physical layer would correspond to the spatial body configuration in terms of position, orientation, and morphology, which intervene as the main factors of transfer functions between the 3D acoustic space and the binaural signals. The Symbolic level would be associated to the listener's knowledge of music material in reference to music theories or cultures, including the way he associates auditory percepts to categories issued from these theories (e.g. pitch quantization into scales). At

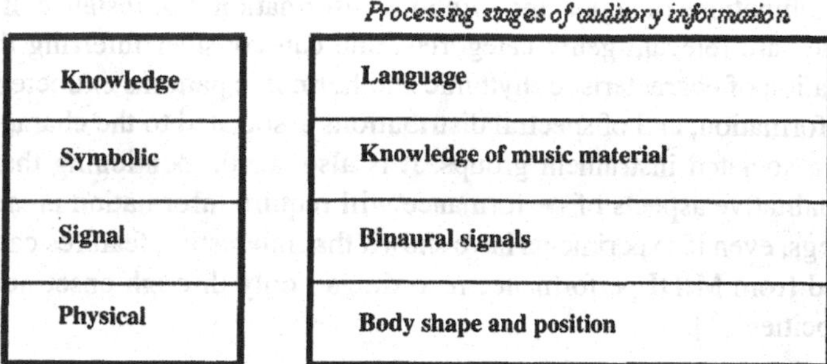

Fig. 2. Hypothetic mapping between representation levels and processing stages of auditory information

the highest level, the knowledge representation is by definition associated to appropriate language structures for describing musical phenomena. This mapping is illustrated in figure 2.

However, this analogy presents some limitations, due to its inadequacy to reflect the structural complexity of the auditory system, in particular at the intermediary levels associated to the spectro-temporal coding of auditory information, and to model the way higher-level information is structured, which remains fairly unknown.

Since these concepts are intended to modeling music representations in technical systems, and do not pretend to model music cognition, a more objective viewpoint for justifying this ordered organization can actually be found in information theory. Increasing levels are obviously associated to a decreasing information quantity and the information reduction operated between successive layers takes the following specific forms:

- from physical to signal levels, the information reduction is essentially of *spatial nature*: the physical layer can be characterized by the acoustic pressure as a function of space and time, whereas the signal level is a function of time, which corresponds to the acoustic pressure signal captured by a microphone at a given space position.
- from signal to symbolic levels, the information reduction is associated to a *double digitization*, which concerns both the *time axis* and the *value ranges* taken by analyzed variables. For instance, in typical signal to MIDI applications, the fundamental frequency is extracted from audio signals as a low bandwidth, slow variation signal, then it is again quantized over time into note events, and the frequency values are mapped into a discrete semitone pitch scale.
- from signal and symbolic to knowledge representations, the information reduction is generally the result on one side of a *global temporal integration* at a more or less global time scale, up to the whole piece duration, and on the other side the *projection of appropriate data value combinations to discrete categories*. Knowledge representations actually describe global characteristics of the music material, including objective descriptions such as the piece name, the music genre, the performing artists, the instruments played, as well as qualitative statements related to performance, sound quality, etc. In specific cases, these characteristics can be inferred

from combinations of signal and symbolic information. For instance, if "rock" and "baroque" are relevant genre categories, one can consider inferring them from a combination of characteristic rhythmic and harmonic patterns extracted from symbolic information, and of spectral distributions associated to the characteristic timbres of associated instrument groups. It is also worth mentioning that characterizing qualitative aspects of performance will require information present in audio recordings, even if experiments have shown that interesting features can already be extracted from MIDI performance recordings , only through onset note positions and velocities [4].

3.1 Mapping of Existing Music Standards into the RL Scale

In order to further specify the RL definitions, let us examine how various music data standards map to the RL scale.

- Audio signal formats. Among mono- or stereophonic signals, the most complete representation, i.e. the one with the biggest information quantity, corresponds to analog signals. Digital audio signal are positioned higher in the RL scale, since they result from a double digitization, in time and amplitude, whose translation in terms of information reduction corresponds to a limitation of bandwidth and dynamics (or signal on noise ratio). The various digital audio formats (e.g. AES-EBU) are ranked in the RL scale according to their information quantity per time unit, i.e. as a function of sampling frequency and entropy of single word coding. Multichannel audio coding formats, such as MADI or ADAT, enable the coding of N independent signals, and it is easy to show that their entropy adds a constant log2N offset to the one of single coded signals. Audio compression formats, such as the ones found in MPEG1 (including the popular mp3 format), MPEG2 and MPEG-AAC, operate a significant information reduction (typically 10 times for mp3) without audible effects in broad sound classes, through the integration of psychoacoustic masking effects. This case illustrates once more the difficulty of setting up a straightforward mapping between the RLs and processing stages in the auditory system.
- Symbolic representation formats. MIDI, the most widely spread standard for symbolic representations, combines the coding of various information types, including note events and continuous controllers, through various channels. Its low bit rate and value representation resolutions (semitone scale, velocities coded in 7 bits, etc.) present many limitations for coding musical events. More recent symbolic representation designs such the one used in IRCAM's OpenMusic composition environment [2], provide better abstractions. First, they define data types through an object-oriented formalism, which enables a better data organization, in particular through multiple inheritance of basic types, shown in Figure 3.

Second, they enable the formalization of rhythms as hierarchical structures through integer decompositions of a given pulsation, whereas symbolic representations found in sequencers only represent time as a linear and flat axis. As another interesting case, music notation formats as used in score editors, such as NIFF, are actually positioned at a higher level in the RL scale. Let aside all graphic layout information, notation formats

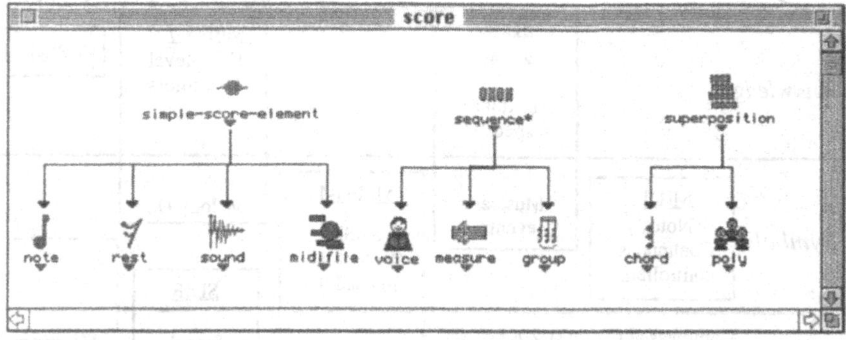

Fig. 3. Basic data types in OpenMusic, from which all music types are derived through multiple inheritance

obviously combine two different kinds of information: on one side, discrete musical events (notes) and specifications (e.g. tempo), which are formalized through numerical data and are thus related to the symbolic level; on the other side, textual qualitative specifications which cannot be translated into numerical data, and are hence part of the knowledge level. So music notation formats are actually positioned in the RLs along the symbolic and knowledge levels.

After targeted efforts, in MPEG1, MPEG2 and AAC, to audio compression techniques, recent evolutions of the MPEG standardization process illustrate the need for integrating new representation levels in audiovisual format standards. MPEG4 enables the representation of compound scenes made of various streamed data including compressed audio [8]; it includes, with SASL (Structured Audio Score Language) a format for specifying musical events; moreover, it also includes, in the Advanced AudioBIFS (BInary Format for Scene description), geometrical and perceptual descriptions of 3D audio scenes. Its position in the RL scale thus extends along the physical, signal and symbolic levels.

The integration of data formats associated to the knowledge level is also developing in various standards, including, for example, UNIMARC for digital libraries, MPEG7 [11], which is dedicated to descriptions of audiovisual contents, and MPEG21, which aims at the identification of audiovisual contents and of their right owners.

The SDIF format[3] (Sound Description Interchange Format), developed by several research centers in music technology, is an open format for representing any kinds of audio analysis data (see 4.2). It fills a gap for such representations both within the signal level and at its boundary with the symbolic level.

The respective positions of these standards in the RL grid is illustrated in Figure 4, which exhibits the lack of a performance-oriented control format which would extend MIDI and enable better representations of gestural control information, as part of the physical level, and mappings of these gestures to symbolic information (cf 3.4.3).

[3] `http://www.ircam.fr/equipes/analyse-synthese/sdif/index.html`

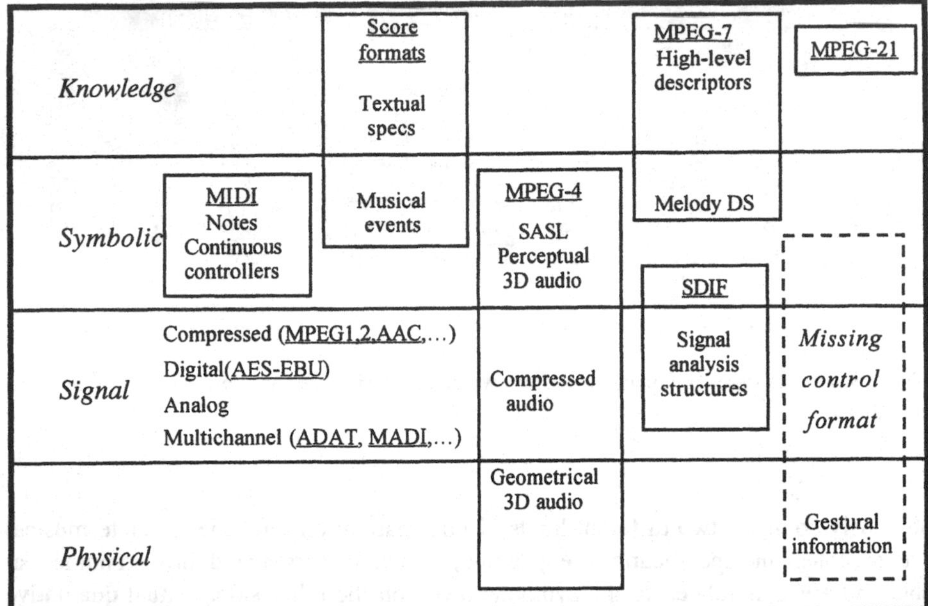

Fig. 4. Mapping of Existing Standards into the RL Scale

4 Analysis of Music Applications through the RL Grid

Since the RL structure is built for providing an extensive view of the various kinds of representations found in music applications, it enables, at least theoretically, to analyze any of them through the grid it provides. This part aims at providing such analysis, through typical examples of music application functions. Therefore, it starts by analyzing usual applications and studying the way they combine the various music representations they use. Then, current research issues in music technology are put into the fore as problems of conversion between various levels and through the integration of the physical and knowledge levels.

4.1 Combination of Various RLs in Usual Musical Applications

Usual musical applications, such as sequencers, score editors, audio processing modules, synthesizers, are generally limited to the management of signal and symbolic representations. Moreover, they manage these representations separately, or with limited interactions. This is true with commercial sequencers, which combine audio and MIDI files. Sequencers provide editing and processing functions for each data type, but with interactions between them mainly limited on one side to the audio rendering of MIDI tracks (further referred as the *Synthetic Performance* function) using synthesizers, on the other side to the overall synchronization of all data on a single time reference, which is the basic function of these applications. As a more sophisticated example, the Max application, which results from research performed at IRCAM on real time interaction

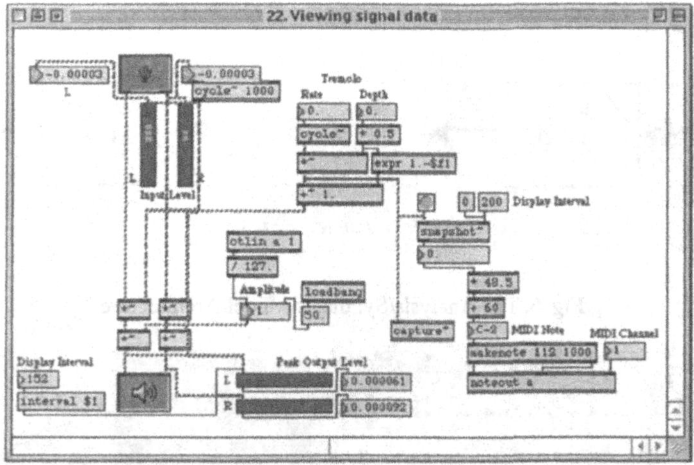

Fig. 5. Example of visual programming patch in Max/MSP. The signal and control processing graphs are respectively represented by dashed and plain lines

[18], provides a dataflow processing architecture specifically designed for managing two kinds of musical information: messages, used in particular for transmitting symbolic information as discrete events, and audio signals. This architecture enables the synchronization of these two data types (Figure 5), through the use of a fixed-size sample block processing architecture for audio signals (typically 64 samples per block), which provides a common clock for all data.

Such an architecture for music information processing, which relies on the synchronization of two clocks, one for audio signals, the other one for discrete and control information, can be actually found in many musical applications. This is the case of DSP plug-ins, i.e. digital audio processing modules designed for commercial sequencer applications.

Functionally, these modules perform the processing of audio signals, through the use of signal processing algorithms, controlled by parameters sampled at a lower frequency rate. This enables to develop sophisticated processing features, which rely not only on signal algorithms, but also on the processing of control parameters. This approach is developed in particular in the GRM Tools processing module suite; each module combines two stages of processing algorithms, the one for audio signal algorithms, the other one for the algorithms parameters, through the systematic use of graphical interfaces [23] and high-level parameter controls. Another conceptual approach of signal processing, developed at IRCAM by Xavier Rodet, relies on an Analysis/Synthesis architecture, through the use of parametric models. Two distinct model categories are considered: signal models, which describe relevant information through signal processing formalisms, and physical models, which provide acoustic models of sound sources and are hence focused on the causes of sound production, whereas signal formalisms model the effects. Depending on the model formalisms, analytical procedures can be associated to synthesis models. In that case, this enables to derive a specific processing architecture, based on the processing of the parameters, as shown in Figure 6.

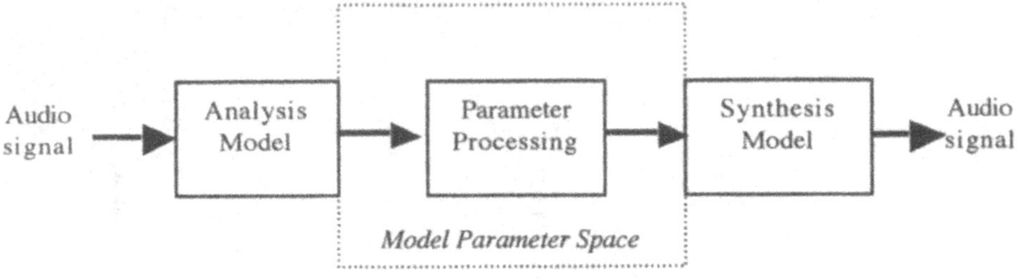

Fig. 6. The Analysis/Synthesis Model Architecture

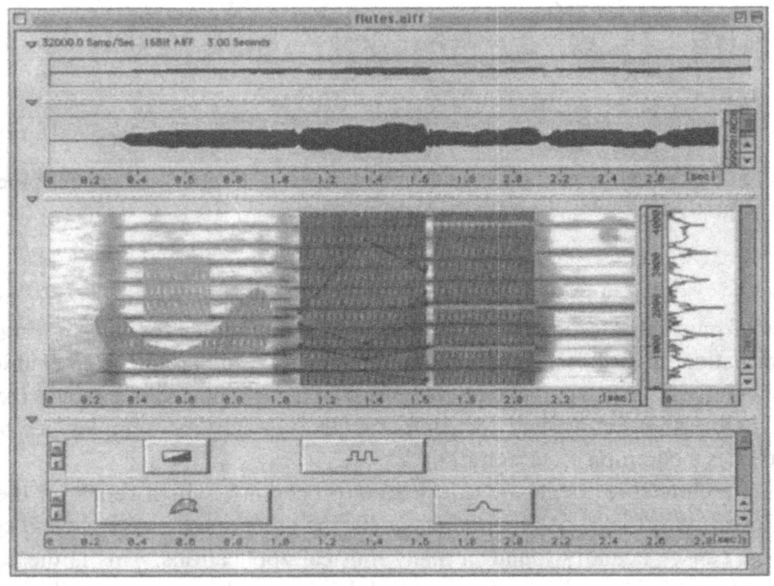

Fig. 7. Time-Frequency Editor of the Audiosculpt 2 Application

This architecture is used in particular in the IRCAM Audiosculpt[4] software application, which is based on the Phase Vocoder model (Short Time Fourier Transform). The processing is specified through a graphical interface which displays the model parameters in a time-frequency graphical representation (sonagram), on which the user can draw in order to define editing operations such as time-varying filters (Figure 7).

In this example, the parameter space (short time Fourier coefficients) actually presents an information quantity of the same magnitude order as the original signal, or even greater. Other models, such as the sinusoidal, or additive model, which decompose an existing signal into a sum of slow-varying sinusoidal functions and a residual signal called noise, enable to have a more quantized parameter space. This space, which includes low-bandwidth amplitude and frequency values of the sinusoidal functions, can be assimilated to the symbolic representation level. In that case, the processing func-

[4] `http://www.ircam.fr/produits/logiciels/audiosculpt-e.html`

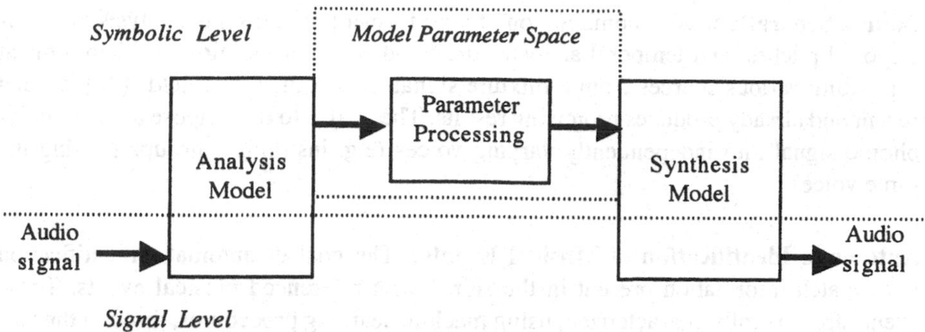

Fig. 8. Audio Processing through Symbolic Information Processing

tion relies on a dual signal to symbolic (analysis) and symbolic to signal (synthesis) conversion, and the processing is specified in the symbolic space (Figure 8).

Such an architecture is widely used in contemporary music productions, which combine recorded audio and instrumental notation materials, through the same symbolic formalism.

4.2 Signal Analysis Issues

The processing scheme of Figure 8 relies on the assumption that an analysis model can be derived from the synthesis model, which is not always the case. Invertible Models, such as the Phase Vocoder, are actually an exception, and deriving analytical procedures from musical signals is currently the subject of many research projects on music technology [6]. Examples of current research issues are listed hereinafter.

Frequency Domain Analysis. Various kinds of information related to the frequency domain can be derived from audio signals: fundamental frequency, spectral peaks, spectral envelopes, etc. In particular, the identification of a limited number of pitch values in a signal mixture provides a digitization from the audio to the symbolic levels. However, in current state of the art in signal analysis, it is still impossible to design robust algorithms capable of analyzing multiple fundamental frequencies present in a polyphonic recording [5].

Automatic Segmentation. The goal of automatic segmentation is to identify time occurrences corresponding to the start and end of musical events. Various time scales can be considered, from the note level (symbolic level) [19] up to the part level in a piece of several minutes (knowledge level) [16]. Some analytical models do not provide only their results as a list of time-stamped events, but also rely on high-level models, such as Hidden Markov Model state sequences: this enables modeling events as compound structures, e.g notes as sequences of various states (Attack, Sustain, Release). However, the unsupervised extraction of musical events from audio signals can be difficult to achieve in some cases. Score alignment algorithms, which perform a synchronization of a reference score expressed in symbolic format with a recorded performance signal, provide more robust results in the general case [15].

Source Separation. As a combination of both former problems, i.e. analysis of superimposed pitches, and temporal analysis, the blind source separation problem aims at separating various sources from a mixture signal. Research in this field [22] is quiet recent and already produces promising results. The goal is to decompose a mixed polyphonic signal into independently varying voices (e.g. instrument groups playing the same voice).

Automatic Identification of Musical Events. The goal of automatic identification is to match information present in the signal with referenced musical events. These events are generally characterized, using machine learning procedures, through the values taken by a vector of numerical descriptors automatically extracted from the signal. An usual form of identification is automatic classification, applied for example to the identification of sound sources (instruments) present in an audio signal, through learned classes characterizing each sound source. Flat and hierarchical automatic classification procedures enable the mapping of identified events to existing taxonomies and can thus be assimilated to automatic conversion functions from the signal level to the knowledge level. New applications, such as digital audio databases, use these automatic indexing features as a computer assistance for classifying sounds in the database [17].

Analysis from Symbolic Representations. Other approaches aim at extracting high-level musical structures from symbolic representations such as MIDI data. Existing research in this field includes automatic meter extraction [20], unsupervised style characterization [1], pattern analysis and matching of melodic, rhythmic and harmonic materials [12]. In the context of music information retrieval, these works, such as the popular query by humming application, generally rely on the analysis of music patterns, between which similarity metrics are applied. One could then wonder where such patterns are positioned in the RL scale, possibly at a missing position between the symbolic and knowledge levels or, like in OpenMusic, as compound structures inherited from basic symbolic objects (notes, chords, event sequences). However, as a specificity of music information, it appears that there is no objective way of producing such patterns from symbolic information; these patterns, such as melodic profiles or surfaces, or rhythmic patterns, only account for a certain aspect of music information, but are not standard elements of the musical syntax. In other words, if compared to language, music includes structural information at the grapheme level (individual notes can be assimilated to individual letters), but not at the lexeme level (there is no structural equivalent to words, as the elementary syntactic and semantic level).

4.3 The Synthetic Performance Problem

We saw in the former examples that a conversion from a low to a higher representation level is done through an analytical procedure, which reduces the information quantity by isolating the appropriate events. Inversely, a conversion between a higher and a lower level will require the generation of additional information, through specific synthesis models. This problem is illustrated through a generic music problem, called here the *Synthetic Performance* application, which aims at simulating the action of a performer and his instrument. In terms of information processing, the performer (plus instrument)

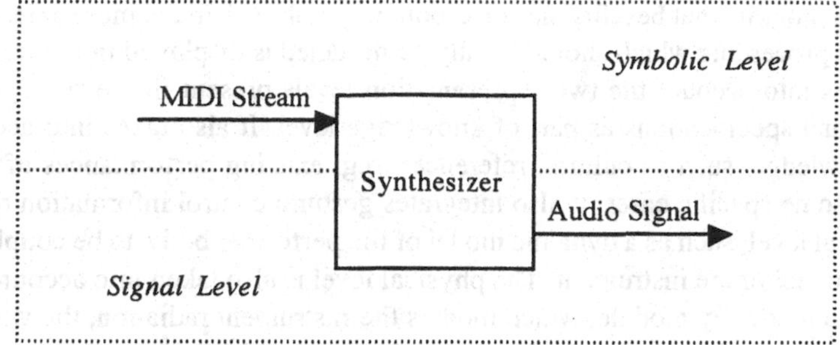

Fig. 9. Basic Performer as MIDI to Audio Function

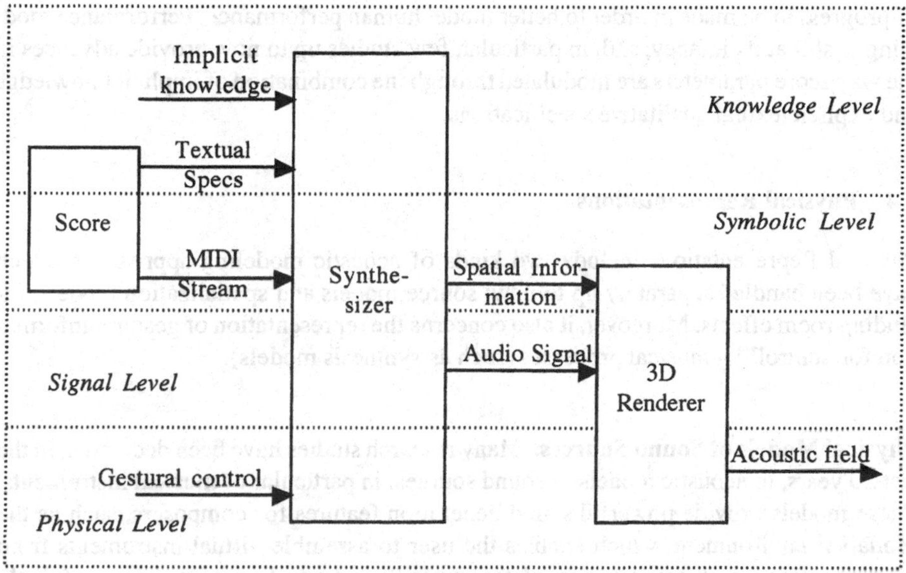

Fig. 10. Ideal Modeling of the Synthetic Performer

function can be summed up as taking symbolic data in input (the score), and producing a variable acoustic pressure at each point of the concert hall space. The computer simulation of this function thus corresponds to a data conversion from the symbolic to the physical levels. A usual way of managing this function in existing applications, generally referred as "MIDI to audio" function, is to assimilate, as shown in Figure 9, the score as a MIDI note sequence, and play it on a synthesizer.

The synthesizer actually performs the symbolic to audio conversion function by taking symbolic information such as pitch, intensity, duration in input, and generating a signal which simulates a note played by a given instrument. The physical level is thus ignored, as well as most score nuances and indications, and the resulting timbre depends on the selected synthesizer program. A more accurate way of rendering for this Score to MIDI function is to use the General MIDI format, which also transmits, from a standardized instrument set, the instrument to be played to the synthesizer, which then selects the program that best fits the corresponding timbre. A much more extensive view of various processing which should ideally be modeled is displayed in Figure 10.

It takes into account the two representation levels present in the score, including some textual specifications as part of knowledge level. It also takes into account implicit knowledge, such as cultural references, e.g. existing performances of the same piece. In some specific cases, it also integrates gestural control information relating to the physical level, such as a dynamic model of the performer body, to be coupled with a physical model of the instrument. The physical level is also taken into account through a 3D audio rendering module, which models the instrument radiation, the way this radiation is controlled by the musical text, the simulation of the concert hall room effect, and finally the rendering of all of these spatial effects on the target reproduction system (headphones, stereo, multiple loudspeakers, etc.). This example illustrates the amount of progress to be made in order to better model human performance. Performance modeling is still at its infancy, and, in particular, few studies up to now provide advances in the way score parameters are modulated through the combination of implicit knowledge and explicit textual qualitative specifications.

4.4 Physical Representations

Physical Representations include two kinds of acoustic modeling approaches which have been handled separately up to now: source models and spatialization models, including room effects. Moreover, it also concerns the representation of gestural information for controlling musical processes (such as synthesis models).

Physical Models of Sound Sources. Many research studies have been dedicated, in the last 20 years, to acoustic models of sound sources, in particular of musical instruments. These models provide powerful sound generation features for composers, such as the Modalys[5] environment, which enables the user to assemble virtual instruments from a set of reference objects (strings, plates, tubes,?) and non-linear interactions (pluck, strike, bow, etc.) [21]. Commercial applications are also developing in the form of real time instrument synthesizers. As compared to signal models, physical models present several advantages: they produce richer and more realistic sounds; their input parameters correspond to physical variables and are thus more meaningful to the user; the simulation of non-linear coupling functions between the various instrument parts enable the reproduction of the instrument behaviour as a non-linear dynamic system, with possible chaotic output depending on input parameter configurations. However, for this reason, physical models are more difficult to control, and present the same kind of difficulties novice performers experiments when starting to play with instruments with a non-linear behaviour such as winds. Therefore, specific research is done focusing on model inversion, i.e. with the aim of producing a given audio signal through appropriate generation of control parameter values over time. Other research projects are also dedicated to the modeling of instrument radiation, and to the way radiation varies according to played notes.

[5] `http://www.ircam.fr/produits/logiciels/modalys-e.html`

Spatialization. Spatialization refers to a simulation function which actually includes two different kinds of parameters that had been handled separately in music technology: localization (sound sources positions), and room effect. Traditionally, these two functions have been simulated respectively using stereo panning (or multi-loudspeaker intensity panning) and artificial reverberation. More recent, physical models enable the complete simulation of the scene from a geometrical description of the room and of the positions of sources and listeners. However, such models, which rely on a convolution of the signal by an impulse response of possibly several seconds, are heavy to compute, and require the impulse response itself to be recomputed as soon as the source or the listener move. Another approach, developed in the IRCAM Spat project [9], is based on a perceptual description of the audio scene. It also combines the localization and room effect simulations, but the room acoustic quality is specified through a set of perceptual parameters. These parameters enable the user to specify the target acoustic quality regardless of the reproduction system, be it headphones (binaural coding), stereophonic (transaural), multi-loudspeakers (intensity panning or Ambisonic), or Wavefield Synthesis. According to the RL grid, the Spat provides a synthesis function from the signal (input sound) and symbolic (target acoustic quality and source positions) levels, to the physical level (reproduction system management). Another, dual approach of spatialization is based on the synthesis of radiation patterns through the control of multi-loudspeaker sources [13]. In the context of live performance and real time processing of acoustic instruments or voice, these sources can be configured in order to approximate the radiation of the processed instrument and thus better fuse with acoustic sounds.

Gestural Control. The generalization of electronic instruments can be considered as a continuity of acoustic instrument building. However, electronic instruments bring a rupture for performers in terms of man-machine interaction: acoustic instruments provide a direct energetic coupling between the gestural control and the produced sound, whereas electronic synthesizers get their energy from the electric power and thus bring a decoupling between the control gesture and the synthesized sound [3]. This enables building various kinds of "new instruments" through the combination of any kinds of gesture capture systems and audio synthesizers, and brings a new issue of defining appropriate mappings between gestural control information and the synthesizer control parameters, addressed by several authors [25]. Moreover, the state of the art in live interaction between performing instruments and real time processing systems relies in score following, i.e. real time music shape recognition from a reference score played live by a performer [21]. In order to go beyond the recognition of symbolic information in terms of expressivity, some composers are interested in integrating gestural control information as input signals of musical processes. These evolutions show the need of appropriate representations of musical gestures, as multidimensional signals resulting from captors or image analysis, characterized by a lower sampling rate than audio (typically 1kHz sampling), but also keeping track of cinematic information such as the trajectory, and position, speed and acceleration as functions of time. Beyond MIDI, there is also a need of a new standard for synthesis control, which would account for various representation levels including symbolic information, but also gestural information.

5 Conclusion

This study has identified four main kinds of music representations in existing and potential applications, and has shown that these four types can be organized in levels associated to the various information quantity they convey. These concepts provide a powerful grid for analyzing all kinds of musical applications according to the types of information they manipulate. They also exhibit the lack of a standard syntactic level corresponding to music patterns, even though such patterns are the only way of characterizing structural information between the note level and the high-level form. These concepts are also useful for understanding current issues in music technology research in terms of integration of the physical and knowledge levels and as problems of data conversion between various levels.From bottom up, analysis functions extract relevant information from complex inputs. Inversely, from top down, synthesis functions generate missing information through dedicated models. In both cases, the various illustrations provided from current research show the gap which remains to be filled between functions that could be envisioned for solving basic musical problems and the current limits of our knowledge.

References

1. Assayag, G. Dubnov, S., Universal Prediction Applied to Stylistic Music Generation., In Mathematics and Music, EMS Diderot Forum 1999, Ed. G. Assayag, J.F. Rodrigues, H. Feichtinger. Springer Verlag (2002).
2. Assayag, G., Rueda, C., Laurson, M., Agon, C., Delerue, O.: Computer-Assisted Composition at IRCAM, From PatchWork to OpenMusic, Computer Music Journal, Volume 23, Number 3, MIT Press (1999) 59-72.
3. Cadoz, C., Continuum énergétique du geste au son, simulation multisensorielle interactive d'objets physiques, In Interfaces homme-machine et creation musicale, Ed H. Vinet and F. Delalande, Hermes Science, Paris (1999), 165-181.
4. Canazza, S., Roda, A., Orio, N., A parametric model of expressiveness in musical performance based on perceptual and acoustic analyses, Proc International Computer Music Conference, ICMA (1999).
5. de Cheveigné, A., Kawahara, H. Multiple period estimation and pitch perception model, IEEE Speech Communication, Vol. 27 (1999), 175-185.
6. Hélie, T., Vergez, C., Lévine, J., Rodet, X., Inversion of a physical model of a trumpet, IEEE CDC: Conference on Decision and Control. Phoenix Arizona (1999).
7. Hiller, L., Music composed with computers, Heinz von Foerster col., J. Wiley & sons, New York (1969).
8. ISO/IEC 14496-1:2000. MPEG-4 Systems standard, 2nd Edition.
9. Jot, J.M., Warusfel, O.: A Real-Time Spatial Sound Processor for Music and Virtual Reality Applications, Proc. International Computer Music Conference, ICMA (1995) 294-295.
10. Mathews, M.V., An acoustic compiler for music and psychoacoustic stimuli, Bell System Technical Journal, 40, (1961), 677-694.
11. Martínez J. M., MPEG-7 Overview, ISO/IEC JTC1/SC29/WG11 N4980, `http://mpeg.telecomitalialab.com/standards/mpeg-7/mpeg-7.htm`.
12. Meudic, B., St James, E., Automatic Extraction of Approximate Repetitions in Polyphonic Midi Files Based on Perceptive Criteria, Computer Music Modelling and Retrieval, LNCS 2771, Springer-Verlag (2003) This issue.

13. Misdariis, N., Nicolas, F., Warusfel, O., Caussé, R.: Radiation control on a multi-loudspeakers device, Proc. International Computer Music Conference, ICMA (2001) 306-309.
14. Orio, N., Lemouton, S., Schwarz, D., Score Following: State of the Art and New Developments, Proc. International Conference on Musical Expression, (NIME-03) (2003), 36-41.
15. Orio N., Schwarz, D., Alignment of Monophonic and Polyphonic Music to a Score, Proc. International Computer Music Conference, ICMA (2001).
16. Peeters, G., La Burthe, A., Rodet, X.,: Toward Automatic Music Audio Summary Generation from Signal Analysis, Proc. International Conference on Music Information Retrieval, IRCAM, Paris (2002).
17. Peeters, G., Rodet, X., Automatically selecting signal descriptors for Sound Classification, Proc. International Computer Music Conference, ICMA (2002).
18. Puckette, M., FTS: A Real-Time Monitor for Multiprocessor Music Synthesis, Computer Music Journal 15(3), MIT Press (1991) 58-68.
19. Rossignol, S., Rodet, X., Soumagne, J., Colette, J.L., Depalle, P., Feature extraction and temporal segmentation of acoustic signals, Proc. International Computer Music Conference, ICMA (1998).
20. Scheirer, E.,D. Tempo and Beat Analysis of Acoustic Musical Signals J. Acoust. Soc. Am. 103:1, (1998), 588-601.
21. Vergez, C., Bensoam, J., Misdariis, N., Caussé, R.: Modalys: Recent work and new axes of research for Modalys, sound synthesis program based in modal representation, Proc. 140th Meeting of the Acoustical Society of America (2000).
22. Vincent, E. Févotte, C., Gribonval, R. and al., A tentative typology of audio source separation tasks. Proc. 4th Symposium on Independent Component Analysis and Blind Source Separation (ICA 2003), Nara, Japan, (2003).
23. Vinet, H., Delalande, F.: Interfaces homme-machine et création musicale, Hermes Science, Paris (1999).
24. Vinet, H., Herrera, P., Pachet, F.: The CUIDADO Project, Proc. International Conference on Music Information Retrieval, IRCAM, Paris (2002) 197-203.
25. Wanderley, M., Battier, M., Trends in Gestural Control of Music, CDROM book, IRCAM, Paris (2000).

“Leçons”: An Approach to a System for Machine Learning, Improvisation and Musical Performance

PerMagnus Lindborg

permagnus@noos.fr
www.notam02.no/~perli

Abstract. This paper aims at describing an approach to the music performance situation as a laboratory for investigating interactivity. I would like to present *“Leçons pour un apprenti sourd-muet”* [1], where the basic idea is that of two improvisers, a saxophonist and a computer, engaged in a series of musical questions and responses. The situation is inspired from the Japanese *shakuhachi* tradition, where imitating the master performer is a prime element in the apprentice's learning process. Through listening and imitation, the computer's responses get closer to that of its master for each turn. In this sense, the computer's playing emanates from the saxophonist's phrases and the interactivity in “Leçons” happens on the level of the composition.

1 Composition Process Formalised

The architecture of “Leçons” takes as point of departure the information exchange between two agents in a conversation-like situation as described in linguistic theory [2]. Communication between a musician and a machine may be considered to have aspects of both natural and formal languages.

As Noam Chomsky states, “In the study of any organism or machine, we may distinguish between the abstract investigation of the principles by which it operates and the study of the physical realization of the processes and components postulated in the abstract investigation.” [3] Certainly, not all procedures of musical production allow themselves to be expressed formally. The composer's (or improvising musician's) mode of operation and level of creativity are neither constant nor continuous, but, allowing simplification, we can outline the borders between three of the sub-processes that are involved: analysis, composition, and diffusion.

In a first phase, the musician internalises exterior stimuli through interpretation and creates models of various degree of formalisation. In the second phase, composition or “putting-together”, aspects of the material are negotiated through experiments, rehearsals and so on. In the third phase, the internal representation is realised as sound, using an instrument. In poetic terms we may call these phases Listening, Composition and Playing. For an improvising musician the situation is similar. Implementing a model of the three processes in a computer program, we isolate two main parts: first a silent part characterised by analysis, and secondly, a sounding part, characterised by production. Linking the two parts is a transformation function, which is internal and may be invisible to an outside observer [4]. The nature of such a transformation is evasive in

U.K. Wiil (Ed.): CMMR 2003, LNCS 2771, pp. 210–216, 2004.

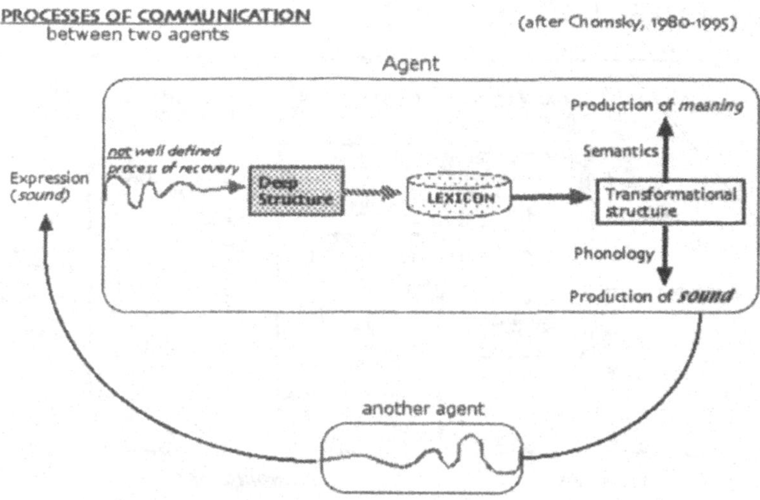

the case of the composition process, and even more obscure in the case of a real-time improvisation performance. Ideally, a computer model to investigate music cognition should be capable of handling large numbers of parameters, of adapting given sets of performance activities to varying situations, and weighing the value of available strategies. "Leçons" in its present state implements a modest approach to the phenomenon of invention. The computer follows a performance script (see below) containing the actions defining a performance. It does not know any notes or phrases and throughout the performance, it listens to the saxophonist, analyses the audio and builds a database of material which allows it to improve its playing. The computer registers aspects of the style of a human improviser. The material is recombined using a statistical method (Markov chains) generating algorithmically its phrase responses (see below) which are played in real-time. The goal of "Leçons" is to construct a situation where the music is created during the performance and where the two improvisers are compatible in terms of producing musically interesting material.

I will now describe how the computer part works. The illustration above shows the three main processes: analysis, recomposition, and synthesis. The program is implemented in MaxMSP [5]. The global form, described in the performance script, is a set of variations, and connects back to the master – apprentice idea. Each variation consists of two parts: first, the saxophonist plays solo while the computer listens; then, the two play together.

2 Audio Analysis and Categorisation

In the Listening part, the computer performs automatic analysis on a continuous audio flow. Required here are methods for data reduction and storage efficiency. The analysis works simultaneously on three different time scales. Going from longer duration to shorter, the data is interpreted as Phrases, Note-categories and Note-models. The high-level musical phenomena are mapped to simple parameters in such a way that they

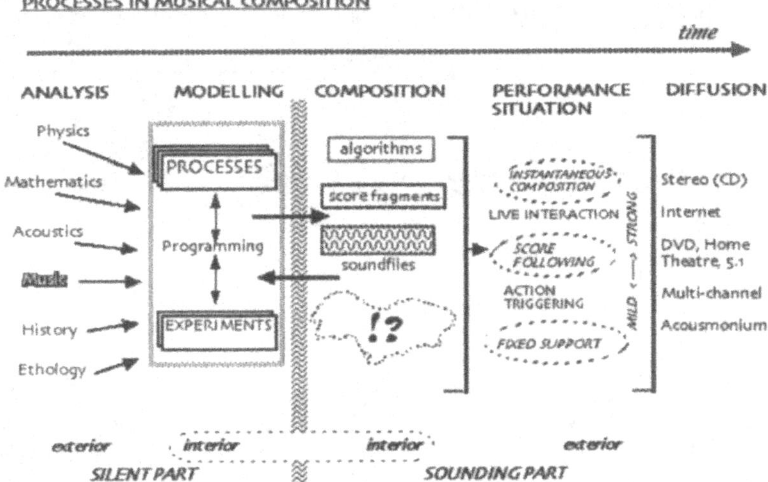

can be manipulated later, in the process of algorithmic composition that generates the computer's phrase responses.

- "Phrases": an amplitude follower picks up the pattern of the sax player's alternations between playing and silence. The database stores an image of the phrase as a series of windows representing these alterations, which can be stretched and applied to any phrase response.
- "Note-categories": the audio stream of the saxophone's melodic playing is chopped up into Notes by detecting pitch [6] and attack (or rather "end-of-silence"). When certain time-related thresholds are passed (these have to be found during rehearsals through trial-and-error) the detection of a new note is triggered. It is defined by its duration, average amplitude and average pitch and these values are approximated in 22 pitch registers, 4 nuances of amplitude, and 7 durations (logarithmic), called "coding classes". The coding class decide into which of the Note-categories[1] the new note is stored, or rather, its (non-approximated) pitch-amplitude-duration values (see detail of the user interface below).
- "Consecutive-note-categories" is a register of the order of appearance of the Note-categories, providing the material for the statistical recombination of detected notes[2].
- The "Note-models" provide a notion of timbre. Over the duration of a detected note, however short, there are fluctuations of pitch and amplitude. These values are sampled and stored as pitch and amplitude curves. The Note-models are later fitted to the notes played by the lead granular synthesis instrument (see below).

---

[1] In this case, since $22 * 7 * 4 = 616$.

[2] During a 10-minute performance, around 600 notes are picked up, so the number of Note-categories seems reasonable.

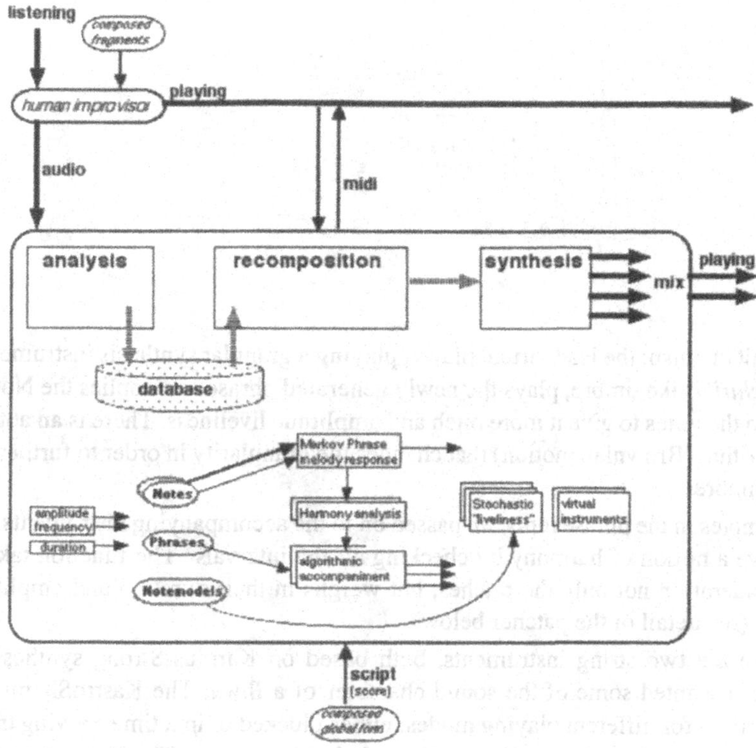

3 Instantaneous Recomposition

In the Playing part, the computer will choose the phrase image (picked up from the saxophonist) that is "richest", i.e. has most changes between sound and silence. The phrase image is fitted to the duration of the section (the remaining part of the variation.) When there is a "play" window in the Phrase, a "melodic phrase maker" is invoked. This is the core algorithm of the system, mapping the database material to play actions [7]. From the data in the Consecutive-note-categories register, a first-order Markov algorithm generates a sequence of Note-categories, which maintains statistically significant properties of the original material. For each of the Note-categories in the new sequence, one of the notes stored in the corresponding slot is chosen at random. The phrase is then passed on to the virtual instruments (see detail of the patcher below).

4 Sound Synthesis

The computer plays an ensemble of four virtual instruments. One takes on a lead role, while the three others are accompanying. The real-life musical metaphor would be that of a group of musicians, something of a cross between a traditional Japanese ensemble and a jazz combo.

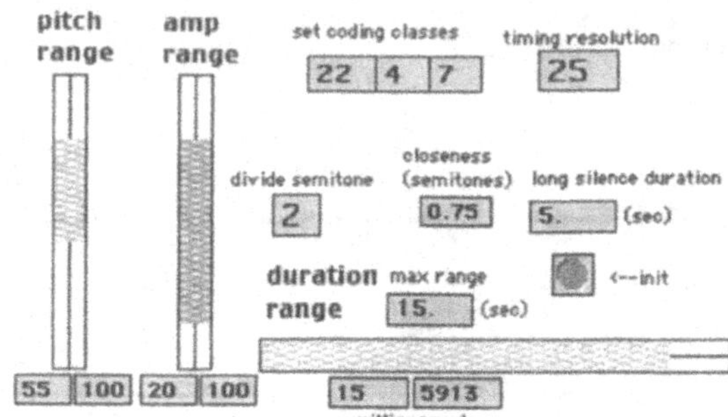

- SashiGulenisu: the lead virtual player, playing a granular synthesis instrument with a *hichiriki*-like timbre, plays the newly generated phrase and applies the Notemodels to the notes to give it more pitch and amplitude liveliness. There is an automatic algorithm (Brownian motion) that changes the granularity in order to further enrich the timbre.

  The notes in the phrase reply are passed on to the accompanying instruments, which derive a notion of harmony by checking all the intervals[3] The function takes into consideration not only the pitches, but weighs in their duration and amplitude as well (see detail of the patcher below).
- There are two string instruments, both based on Karplus-Strong synthesis [9]. Here, I wanted some of the sound character of a *Biwa*. The KastroShami has an algorithm for different playing modes: either plucked or in a time-varying tremolo. In either case, it may introduce a glissando between notes. The KastroKoord plays three-note chords.
- The MeksiBass [10] uses algorithms to create something of a jazz style walking-base. Pitches harmonically closer to the fundamental are favoured, in order to strengthen the harmonic unity of the virtual ensemble.

5 Performance Script

The performance script is the equivalent of a musical score. As an example of what the script looks like, consider the excerpt shown below. The fifth variationt has a total duration of 35 seconds (Listening and Playing parts together); all four instruments except MeksiBass will play; the KastroShami will play very slow-moving phrases (3 times longer than the others) ; the Markov phrase generator will use the most recent forth of the detected notes (the "memory area" is set to 90 120; having access to all the notes would be 0 127).

[5, movementDuration 35 \, instrumentsActivity 1 0 3 1 \, setMemoryArea 90 120]

[3] Algorithm based on [8].

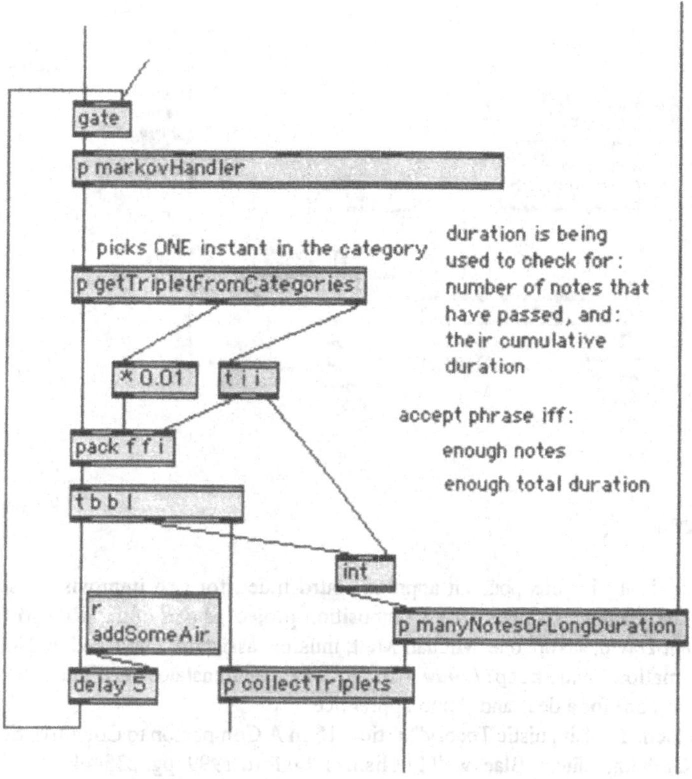

6 Conclusion

We may now consider the performance from a listener's perspective. I refer to the sound examples (1: beginning, 2: towards end) taken from a live recording, When setting out, the computer knows nothing and will perform hesitantly. As it gathers more material, the playing is considerably enriched. "Leçons" implements a procedural system, without attaching musical sense to the data it perceives [11]. When considering the nature of the musical interactivity, the role of the computer is determinant. In an *Automatic* system the machine consists of algorithms for the generation of music, and local unpredictability is often the wanted result. In a *Reactive* system, the computer is treated as an instrument to be performed on, and emphasis is on user interface and direct mapping of controller gesture to sound production. In an *Interactive* system there is a two-way exchange between musician and machine of information that transforms surface elements and deeper structures of a performance in real time [12]. The interactivity in "Leçons" is strong in that the composition is decided during the performance, resulting from an automatic analysis of audio. The interactivity is automatic in that it relies on (prefabricated) algorithms to make musical sense. The computer improviser is not a reactive instrument, and the saxophonist needs to invest time to explore the kind of playing which works within the system's technical limitations and ubiquitous musical setting.

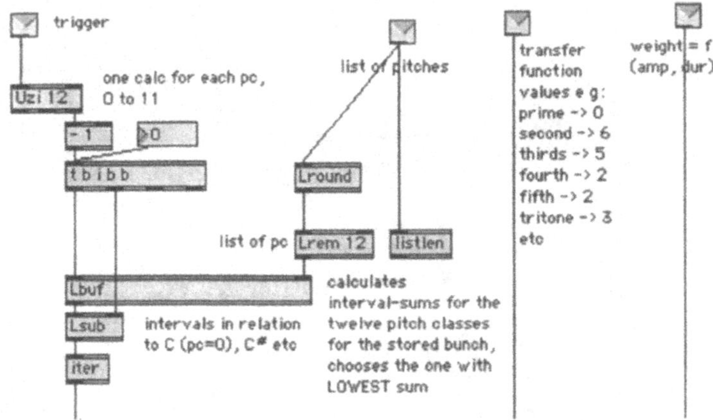

References

1. Lindborg, P.M.: Leçons pour un apprenti sourd-muet, for two improvisers: soprano saxophonist and computer. (1999, 10'). Composition project as part of the IRCAM Cursus 1998-9. Vincent David, saxophone, Michail Malt, musical assistant. Published by Norwegian Music Information Centre `http://www.mic.no`. The title translates into "music lessons (or "the sound", le son) for a deaf-and-dumb apprentice".
2. Langendoen, T.: "Linguistic Theory", article 15 in A Companion to Cognitive Science. Bechtel and Graham, editors. Blackwell Publishers, Oxford 1999, pg. 235-44.
3. Chomsky, N.:Rules and Representations. Columbia University Press, New York 1980, Quote from pg. 226.
4. McCauley, S.:"Levels of explanation and cognitive architectures", article 48 in A Companion to Cognitive Science, id., pg. 611-24.
5. Puckette, M., Zicarelli, D. : MaxMSP: `http://www.cycling74.com`
6. Puckette, M.: "Real-time audio analysis tools for Pd and MSP", Proceedings of ICMC 1998.
7. Russel, S. & Norvig, P.:Artificial Intelligence, a Modern Approach. Prentice-Hall International, Inc. Upper Saddle River, NJ. 1995, pg. 34.
8. Murail, T: "Esquisse Library" for OpenMusic. IRCAM, Paris, 1997.
9. Serafin, S.:"kastro " External object for MaxMSP, `http://www-ccrma.stanford.edu/~serafin/1999`.
10. Msallam, R., Delmar, O., Mårtensson, P. & Poletti, M.: "MEKS: Modalys to Extended Karplus-Strong." IRCAM 1998.
11. Cardon, A.: Conscience artificielle et systèmes adaptatifs. Eyrolles, Paris, 1999, pg. 281-3.
12. Lindborg, P.M.: Le dialogue musicien-machine: Aspects de systèmes d'interactivité. Mémoire de DEA, Université de Paris-4 Sorbonne 2003. `http://www.notam02.no/~perli`

About the Role of Mapping in Gesture-Controlled Live Computer Music

Leonello Tarabella[1] and Graziano Bertini[2]

ComputerART project of C.N.R
Istituto di Scienza e Tecnologia dell'Informazione "A.Faedo"
Area della Ricerca di Pisa, Via Moruzzi, 1 56126 Pisa, Italy
l.tarabella@cnuce.cnr.it
g.bertini@ei.pi.cnr.it
http://www.cnuce.pi.cnr.it/tarabella/cART.html

Abstract. Interactive computer music proposes a number of considerations about what the audience experiences in relationship of what-is-going-on-on-stage and the overall musical result. While a traditional music instrument is a compact tool and "to play an instrument" has a precise meaning for everybody, the new electro-acoustic instrument is a system consisting of a number of spread out components: sensors and controllers, computer and electronic sound generators, amplifiers and loudspeakers. How to link information between the various parts of this exploded instrument is deeply correlated to new modalities of composing and performing in relationship with how the audience perceives and accepts these new paradigm. We here report our point of view and considerations about the role of "mapping" derived from our experience both in developing original controllers and in the realization of interactive electro-acoustic performances.

1 Introduction

A traditional music instrument is a compact tool which gathers together all the aspects (shape, ergonomics, mechanics and material) necessary for stating and determining the timbre and for controlling pitch and nuances of sound. For some instruments it is also possible to personalize the acoustic response by choosing specific interchangeable crucial elements: mouthpiece (and reeds) for wind instruments, material and size of strings for string instruments. Besides, the physical structure of the instruments reflects the alphabet and syntax of reference for the music played (which in the case of western music is the well tempered scale and Harmony) and reflects even the anatomic structure of the human body.

A question now arises: is the term "instrument" still appropriate and correct for the new equipment used in computer music? Compared to a traditional compact musical instrument the new one appears as an "exploded instrument" consisting of different elements: controller(s), audio-signal generator (the computer) and sound sources (loudspeakers) connected via different typologies of cables and signals.

U.K. Wiil (Ed.): CMMR 2003, LNCS 2771, pp. 217–224, 2004.

There exist two main types of connections: the digital connection between controllers and computer and the analog connection between computer and loudspeakers. The analog path is related to rendering problems: regarding how to enjoy a video-clip or a film, it makes great difference whether using a simple home one-small-loudspeaker TV-set or a Stereo-surround equipment since it affects the quality of sound and fidelity to the original design together with the intentions of the composer and the players.

The digital connections are more crucial. Controllers, or gesture recognition devices, produce data-flows used by the computer for producing sound [1]. The problem now consists in how to link, or better, how to *map* information coming from controllers to programs which generate complex musical structures and/or to synthesis algorithms which generate sound signals.

1.1 The New Instrument

The composer/performer sets up a software mechanism which uses data coming from a controller to produce sound events. The performer "plays"... not precisely an instrument but rather a dynamic meta-instrument: this is the real novelty introduced by music performance. From the point of view of the audience things become more difficult to understand especially when original controllers based on different kinds of sensors (pressure, acceleration, heat, infra-red beams, ultrasound, etc.) or gesture recognition systems based on realtime analysis of video captured images, are used by the performer.

In the computer music field a great variety of very sophisticated and complex gesture interfaces have been designed and realized using almost any kind of sensor [2–9].

From our experience, in particular regarding impressions and questions coming from the audience after our concerts, we argue that people usually can appreciate and understand that what is going on musically comes from the presence and the movements of the performer, but in general are unable to understand the complex cause-effect relationships and usually think the controller is the instrument. And usually the audience is completely unaware about the crucial role of mapping and what actually the computer does during the performance for generating events in accordance with predefined music/acoustic material combined with information coming from the controllers a performer is acting on. The simple one-to-one mapping rule valid for traditional instruments leaves room for a theoretically infinite range of mapping rules definable by the composer for a specific piece and even for each part of that piece. The mapping is a part of the composition.

This approach has open a complete new and wide territory to explore for composition, and especially, for live performance. It is no longer a matter of playing an instrument in the traditional sense but, rather, playing a specific piece of music in terms of activating and controlling during the live performance musical/acoustic material and algorithms prepared during the compositional phase [10, 11].

2 The Mapping Paradigm as a Creative Tool

We shall give here some very basic ideas we usually take into consideration when using mapping as an aesthetic and creative tool for live gesture-controlled computer music performances. We think it's hard to formalize rules and/or strategies about mapping since we are here facing the realm of creativeness and it appears rather difficult to try to follow and/or fulfill a specific syntax while composing. Anyway we give here an informal but usefull definition of mapping as "the possibility of implementing algorithmic mechanisms which dynamically put in relationship data coming from gesture recognition devices and algorithms which generate musical events and sound".

Consider a very simple example where the mouse is used as a gestural interface and a MAX-MSP [12] patch generates sound in accordance with these simple rules: vertical position of mouse sets pitch of sound, horizontal position controls harmonics content, button-down starts sound, button-released stops sound. Another situation could be: pitch is random, timbre is fixed in advance, vertical position controls the attack time, horizontal position controls the amount of reverberation. A further situation maybe ... maybe you, the reader, at this point has devised some different and smarter ideas.

As a consequence of the many ideas and arrangements one can think of how to link the simple and standard functionality of the mouse, we can claim that mapping and composition make part of the same creative activity at both micro level of timbre and macro level of musical melodic and rhythmic patterns.

In [13] Silviane Sapir wrote that "mapping should be neither too simple nor too complex since in the first case the real power of the computer turns out to be not so well used; in the second case the audience is not able to understand what is happening and cannot appreciate completely the artistic content of the performance". Having direct experience of that, we strongly agree with this observation and, further, we think the rule can and must be extended as follows: we experienced that if a complex mapping situation is reached after a growing-up complexity started using simple (close to one-to-one) mapping, the audience willingly accepts it even if highly complex to be understood. It's important however that the "training" phase has a *per sé* aesthetical and musical meaning.

After one or two episodes like that, it is possible to use the opposite path, that is from a very complex mapping situation to a simple one. This will be accepted by the audience because in some way people are faithful that something will happen to "explain" (artistically speaking) what is going on; often it happens that someone starts the guess-the-riddle game in his/her mind. And after a number of episodes like those described, also sharp changes from simple to complex and vice-versa, mapping proves to be of interest and well accepted by the audience.

3 The Importance of the Audience

For us the mapping paradigm is the real novelty in live performed computer music. For that we take the audience into great consideration as the opposite pole of the composer/performer.

In an avant-garde concert executed with traditional musical instruments, the audience is requested to understand and taste the musical language and musical content proposed. A default for the audience is that musicians play musical instruments, that is that they use well known mechanical "tools" for producing sound, in the same manner a speaker is expected to use his/her mouth: attention is focused on the content. In a tape-electronic music concert, the artistic message is accepted as an opera prepared in studio, in the same manner as a film or a video-clip, no matter how the composer reached the result.

But in a live computer music concert the visual component is of great importance when the new "exploded" instrument is used, because attention of the audience is also focused on the relationships between gesture of the performer and the music they are listening to. And people want to understand the rules of the new game, beside tasting and appreciating the musical result.

It is important then to plan a storyboard of different situations each one characterized by well defined musical-acoustic micro-worlds inside of which well balanced amounts of simple and complex mapping arrangements between gesture and music are used.

Our attention shifts now to technical problems and proper solutions related to gesture recognition systems and to mapping.

4 Mapping and Acoustic Feedback

In gesture controlled electro-acoustic musical performances a big role is played by the psychoacoustic feedback, that is the loop created by the performer's movements upon the controllers and the generated sound the performer hears [14]. In traditional wind and bow instruments, feedback is related to the continuous control of sound characteristics (pitch, intensity, timbre, articulation etc.) during the generation by means of continuous modifications of the physical synthesis parameters. The importance of feedback can be easily experienced when it lacks i.e. by playing an instrument with the ears closed or while wearing headphones playing different sounds or music at an high volume; in these cases the intonation and the timbre result differently from those desired because even tough the movements and postures of the body onto the instrument are close to the "correct" due physical values, little parameter differences cause audible sound differences.

Acoustic feedback is equally important in realtime controlled computer generated music; however, as seen before, the "new instrument" entails a completly new behaviour due to the number and the typology of elements involved. Actually, since the new instrument is indeed a system, knowledge about System Theory [15] can be applied for a pertinent investigation and usage of the input and output data-flow. In this field we know that the typical concepts to take into consideration are: *instability, controllability* , *linearity* and, in presence of digital devices, *sampling rate, quantization, latency and multiple triggering.*

4.1 Instability and Low Controllability

Instability means that a system under finite stimuli produces an infinite and non-decreasing response. Controllability indicates to what extent it's simple or not to control the system states and the output by varying its input. Controllability can be low or high: low controllability, which for traditional instruments could be translated into "difficult to play", typically consists of bad features in the direct path, that is: performer→ controller→ sound; if present, it will appear and will be heard by the performer whenever the instrument is played (reverse path: performer ← sound).

4.2 Linearity

In many kind of controllers most of the sensors used are typically not linear. But after all non-linearity is present also in traditional musical instruments even if not known in these terms: in the violin it is much more difficult to get the correct pitch when the finger gets closer to the bridge due to the non-linear response of the pitch versus the finger position?and no violin player complains of that. Anyway when non-linearity is a problem, proper methodos can be used for linearization using mapping, otherwise, as it happens for volume or pitch controls, values can be directly used. Both behaviours can be avoided or used depending upon the artistic and creative needs: for example, discontinuity should be implemented in mapping with a "threshold crossing approach" when the desired output is a trigger and the input is "continuous".

4.3 Sampling and Quantization

We can assume that all gesture controlled musical systems have a digital part; in order to convert analog into digital signals we know it is necessary to use two types of processing: sampling and quantization also called Analog-to-Digital conversion (A/D). The gesture capturing systems have low sampling rates, about some ten Hertz. If we try to directly control low level sound synthesis with such a low rate signal we will hear a lot of clicks; some precautions must then be taken into consideration in order to avoid them.

The second step in A/D conversion is quantization, where a finite and limited number of bits are to be used, typically 8 or 10, for representing values coming from controllers. Even in this case it is usually unwise to directly control low level synthesis parameters, since the "steps" in sound signals can be heard, especially when controlling the pitch.

Oversampling and related interpolation techniques are used to solve both of the above mentioned problems in order to increase the time and the amplitude resolutions [16].

The resulting signal is "more continuous" or, better, "less discrete" from a practical point of view since it amounts to an higher sampling frequency and uses a greater number of bits than the original. When necessary (for example in pitch or timbre variations) it's so possible to control sound synthesis without audible clicks and steps.

4.4 Latency

This is a well known concept in the computer music field and it is generally defined as "the delay between the stimulus and the response".

While in traditional instruments there is usually no latency since the effect (sound) is emitted as soon as the stimulus (bow movement, key hit, etc) is started, in the new instrument two types of latency are present: the short time latency and the long time latency [17]. Short time latency (10 ms order of magnitude) depends on the audio-signal buffers size, on the sampling rate and on the different kind of data processing; this is always present in digital processing systems. When the delay between cause and effect is too high, the response is perceived in late and both the performer and the audience realize that the system does not *respond* promptly.

On the other hand, the long time latency can be used as a specific compositional tool as many composers do for implementing specific sound effects or data processing.

4.5 Triggering

Another important point for mapping, especially when triggering sound samples, is the anti-bouncing algorithms. When a sound sample is triggered from a signal coming from the gestural recognition device, it can happen that instead of only one single trigger several of them come one after another. In this case the multiple triggering, if not filtered, will make the sample start many times and lots of "clicks" will be heard at audio level. In order to avoid multiple triggering it is necessary to filter out the triggering signal once the sample is started for a time depending upon the sample duration. This problem, called *synchronisation*, does not appear in musical interfaces only, but it is typical of the interfacing between analog and digital circuits; anti-bouncing hardware or algorithms are always implemented in the keyboards of calculators, computers, mobile phones etc..

While in specific technical application all these problems must be taken into account and must usually be solved in order that they work properly, in the creative artistic context the composer/performed is requested to be aware of them; they should be taken into consideration but it is not strongly requested to solve them since if sometimes they can cause unwanted results, at other times can be used for reaching specific artistic goals and often they must even be emphasized.

5 Conslusion

In this article we focused attention on mapping from three different and complementary points of view and approaches: philosophical, technological and artistic. As said, it's not a matter of formalizing mapping but, rather, it's a matter of being aware as much as possible about the features mapping offers for expressive/artistic purposes. Since mapping also leaves space for improvisation [18] the presence of the audience is extremely important for its usage as a new tool for making music. And direct human-to-human artistic communication gives back useful information for that.

The MAX and MAX/MSP languages allows the philosophy of mapping as a new territory for creative activity to be put to work. While Max is a de facto standard, at the moment there do not exist standards for gesture tracking systems and it seems that the activity of designing and carrying out personal and original interfaces is particularly rich as that of composing and performing [18].

Technology *per sé* is not enough for novelty: from the iconography point of view there are not crucial differences between a rockstar singing and playing an electric guitar on stage using the most recent and sophisticated electronic equipment and a renaissance or medieval *menestrello* playing his lute and singing love and war songs. Mapping could be the true novelty.

We hope our considerations and results may be of some utility to some beginner in this fascinating territory of creative music.

Acknowledgements

Special thanks are due to Massimo Magrini who greatly contributed in developing the proper communication and audio synthesis software and the true implementation of gesture tracking systems at the cART project of CNR, Pisa.

References

1. Tarabella, L., Bertini, G., Boschi, G.: A data streaming based controller for real-time computer generated music. *In Procs of ISMA2001 (International Symposium on Musical Acoustics)*, Perugia, Italy (2001) 619–622.
2. Tarabella, L. Bertini, G.: Wireless technology in gesture controlled computer generated music. *In Procs of the Workshop on Current Research Directions in Computer Music, Audiovisual Institute*: Pompeu Fabra University Press., Barcelona, Spain (2001)
3. Wanderly, M: `http://www.ircam.fr/equipes/analyse-synthese/wanderle/Gestes/Externe/`
4. Camurri, A., Hashimoto, S., Ricchetti, M., Ricci, A., Suzuky, K., Trocca, R., Volpe, G.: Eyesweb: Toward Gesture and Affect Recognition in Interactive Dance and Music Systems: *Computer Music Journal Spring 2000, 24:1.* (2000) 57–69.
5. Paradiso, J.: Electronic Music: New Ways to Play: *IEEE Spectrum Computer Society Press. Dec. 1997.* pp. 18–30.
6. Povall, R.: Realtime control of audio and video through physical motion: Steim's bigeye: *Proceedings of Journées d'Informatique Musicale* (1996)
7. Tarabella, L., Magrini M., Scapellato G.: Devices for interactive computer music and computer graphics performances: *In Procs of IEEE First Workshop on Multimedia Signal Processing,*Princeton, NJ, USA - IEEE cat.n.97TH8256 (1997)
8. Serra, X.*Proceedings of the MOSART (HPRN-CT-2000-00115) Workshop on Current Research Directions in Computer Music,* Audiovisual Institute, Pompeu Fabra University, Barcelona, Spain. (2001)

9. Mulder, A.: Virtual musical instruments: Accessing the sound synthesis universe as a performer. *In Proc. of the First Brazilian Symposium on Computer Music.* (1994)
10. Rowe, R.: *Machine Musicianship.* Cambridge. MIT Press. March (2001)
11. Tarabella, L., Bertini G.: Giving expression to multimedia performances; *ACM Multimedia, Workshop "Bridging the Gap: Bringing Together New Media Artists and Multimedia Technologists"* Los Angeles, (2000).
12. Winkler, T.:*Composing Interactive Music, techniques and ideas Using Max.*The MIT Press (1999)
13. Sapir, S.: Gesture Control of Digital Audio Envirinments. *In Journal of New Music Research, 2002, Vol.31, No 2, pp.119,129,*Swets & Zeitlinger
14. O'Modhrain, M. S.: *Playing by Feel: Incorporating Haptic Feedback into Computer-Based musical Instruments.*Ph.D. Dissertation, Stanford University Press. (2000)
15. Marro, G.:*Controlli Automatici:* Zanichelli Press, Italy (1997)
16. Oppenheim, A.V., Schafer, R.W.: *Digital Signal Processing.* Prentice Hall Press. (1975)
17. Wessel, D. Wright, M.: Problems and Prospects for Intimate Musical Control of Computers. *In Proc. of the 2001 Conference on New Instruments for Musical Expression (NIME-01),* Seattle. (2001)
18. Tarabella, L.: Improvvisazione Jazz e Performance Interattiva con l'uso dello strumento informatico. *In Proc. of the 2nd Int. Conference on Acoustics and Musical Research.*Ferrara. Italy. Pedrielli Press. (1995)

Designing Musical Interfaces with Composition in Mind

Pascal Gobin[1,2], Richard Kronland-Martinet[3], Guy-André Lagesse[2], Thierry Voinier[3], and Sølvi Ystad[3]

[1] Conservatoire National de Région
2, Place Carli
F-13001 Marseille France
pgobin@wanadoo.fr
[2] Les Pas Perdus
10, Rue Sainte Victorine
F-13003 Marseille France
http://www.unbonmoment.com
[3] C.N.R.S. (Centre National de la Recherche Scientifique)
Laboratoire de Mécanique et d'Acoustique
31, Chemin Joseph Aiguier
F-13402 Marseille France
{kronland,voinier,ystad}@lma.cnrs-mrs.fr

Abstract. This paper addresses three different strategies to map real-time synthesis of sounds and controller events. The design of the corresponding interfaces takes into account both the artistic goals and the expressive capabilities of these new instruments. Common to all these cases, as to traditional instruments, is the fact that their specificity influence the music which is written for them. This means that the composition already starts with the construction of the interface. As a first approach, synthesis models are piloted by completely new interfaces, leading to "sound sculpting machines". An example of sound transformations using the Radio Baton illustrates this concept. The second approach consists in making interfaces that are adapted to the gestures already acquired by the performers. Two examples are treated in this case: the extension of a traditional instrument and the design of interfaces for disabled performers. The third approach uses external events such as natural phenomena to influence a synthesis model. The Cosmophone, which associates sound events to the flux of cosmic rays, illustrates this concept.

1 Introduction

Nowadays computers can make sounds and musical sequences evolve in real-time. Musical interpretation make sense when such new digital instruments offer possibilities of expression. Unlike traditional instruments, an interface based on a digital device is not restrained by the mechanics of the instrument. These new interfaces naturally prepare for new gestures aiming at exploiting, as efficiently as possible, the possibilities offered by the computer. Nevertheless, these huge

U.K. Wiil (Ed.): CMMR 2003, LNCS 2771, pp. 225–246, 2004.

possibilities of sound piloting can sometimes slow down the creative processes if they are not clearly thought through. Even worse, these new technologies might make the technician more valuable than the composer if the technology is used to impress the audience ignoring the importance of creativity and musical intention. This can be the case if the interfaces are made without reflections about the musical context in which it is to be used. In our opinion, the design of an interface is already part of the creative and the compositional processes. In this paper we present four examples of the design of sound interfaces based on three different strategies: the piloting of synthesis parameters to perform intimate transformations on the sounds, the adaptation of interfaces to specific gestures, and the construction of controllers piloted by natural events. This work is the result of an active collaboration between scientists and musicians. It has mainly involved the group S2M (Synthèse Sonore et Modélisation) of the Laboratoire de Mécanique et d'Acoustique, the Conservatoire National de Région and the association Les Pas Perdus all of which are located in Marseille (France). The first example is an attempt to control sound synthesis models by giving the musician a set of new tools to sculpt sounds. These new sound possibilities naturally ask for new interfaces. We found that the Radio Baton, initially designed by Max Matthews to conduct a musical sequence, can be used with great effect to pilot synthesis model parameters. Here the aim was to give the musician an intuitive and easy-to-use tool for improvised sound transformations. The second application points at a quite different issue. Here the problem is to give the musician the possibility of expanding his or her own instrument. Actually, even though a large number of new interfaces have been made, the most common digital instruments use keyboard interfaces. Even though these interfaces offer a lot of possibilities, the musical interpretation can not be the same as the interpretation obtained when playing a wind instrument or a string instrument, since the instrumental play is closely related to the physics of the instrument. This means that for example the linear structure of a keyboard is not easily adapted to the playing of a trumpet, and that the information given by a keyboard is poor compared to the possibilities that a flute player has when playing a sound. Several MIDI controllers with structures close to traditional instruments like wind or string instruments (Machover, 1992) have been proposed. In most cases these controllers are quite limited, since the sensors generally don't give access to fine playing information (for example the lack of reed vibration information in the Yamaha WX7) and are not dedicated to a given synthesis model allowing for example natural sound transformations. Here we show how an interface based on a traditional flute equipped with sensors and a microphone would give the performer access to synthesized sounds through flute-like synthesis models, without modifying the traditional playing techniques obtained through years of practicing. Another example of interfaces suited to special gestures addresses the design of controllers adapted to a set of unconventional motions. This is an artistic project, which has been going on for about five years, with the participation of four people with limited gestural faculties. The main features of this work lie in the fact that the technological realization is closely linked to the artistic problems and the heavy conditions of the handicap itself. In particular, rather than setting ourselves a target leading to a conventional situation (e.g. 'playing music'), and of imagining with the handicapped people the necessary technological methods with this target in view, we preferred to start from scratch and to adapt completely new tools for them. As we already mentioned, the last group of interfaces connects sound models to natural phenomena. This is a degenerate case where the "motion" can not even be learnt by the "player" but has to be taken as it is. The last example, which illustrates this approach, has been made possible thanks to collaboration with Claude Vallée and David Calvet of the Centre de Physique des Particules de Marseille. In this case a device, which we have called the Cosmophone has been designed to make the flux and properties of cosmic rays directly perceptible within a three dimensional space. This is done by coupling a set of elementary particle detectors to an array of loudspeakers by a real time data acquisition system and a real time sound synthesis system. Even thought the aim of the original installation was to make people aware of a physical phenomenon in an entertaining way, such a system can also be used in a musical context.

2 Sculpting the Sounds Using the Radio Baton

The conception of sensors detecting movements is an important part of Max Mathews' current research work. One of his inventions, the Sequential Drum, consists of a surface equipped with sensors. When an impact is detected on this surface, the apparatus indicates the intensity and the position of the impact on the surface (Mathews and Abbot, 1980). A collaboration with Boie (Boie et al. 1989) led to the design of the Radio Drum which detects the motion even when there is no contact between the surface and the emitters. The Radio Drum is in fact able to continuously detect the position of the extremities of the two drumsticks (emitters). These first prototypes were connected to a computer containing an acquisition card piloted by a conductor program allowing a real-time control of the execution tempo of a partition already memorized by the computer (Mathews 1991a, 1991b 1997). The Radio Drum is also called the Radio Baton, since the drumsticks launching musical sequences can be used in the same way as a baton used by a conductor of an orchestra. Max Mathews designed the Radio Drum for two reasons:

- to make it possible to actively listen to a musical play by releasing musical sequences and thus give a personal interpretation of a musical work.
- to make it possible for a singer to control his or her own accompaniment.

The Radio Drum comprises two sticks (batons) and a receiver (a 55 x 45 x 7 cm parallelepiped) containing the electronic parts. In a simplified way, each drumstick can be considered as an antenna emitting radio frequency waves. A network with five receiving antennas is placed inside the drum. It measures the coordinates of the placement of the extremities of the drumsticks where the emitters are placed. The intensity of the signal received by one of the antennas depends

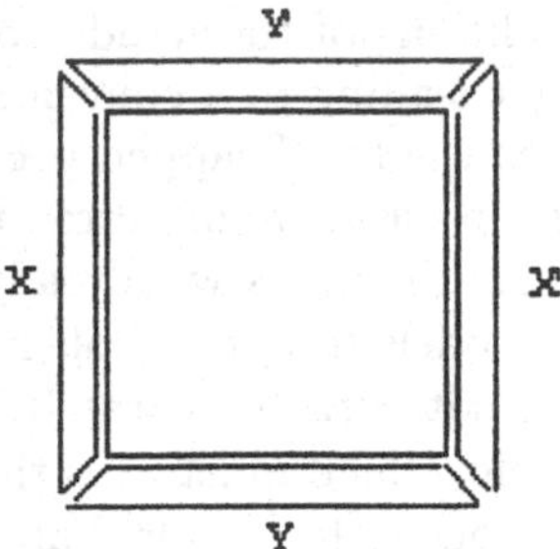

Fig. 1. Placement of the receiving antennas of the Radio Baton.

on its distance from the emitter: it gets stronger as the emitter gets closer to the antenna. To evaluate the position of the extremity of the baton as a function of the x-axis, it is sufficient to calculate the difference in intensity between the antennas x and x' (1). In the same way the antennas y and y' give the position of the emitter as a function of the y-axis. In order to get information about the height of the baton (the z-coordinate), it is sufficient to add the intensities received by the five antennas. Since each baton emits signals with different frequencies, it is relatively easy to discriminate between the signals sent by the two batons to each antenna. The relation between the coordinates and the intensity of the received signals is not linear. A processing of these data is necessary so that the system gives information proportional to the coordinates of the batons. This operation is realized by a microprocessor making it possible to add other functions to the instrument. The latest version contains additional software either making it possible to transmit the coordinates of the batons (and information about the controllers) when requested from a computer, or to transmit the coordinates of one of the batons when it cuts a virtual plane parallel to the surface of the drum the same way as when one hits its surface. When the program detects the virtual plane, it calculates and transmits the velocity of the movement as a function of the z-axis. The height of the plane can of course be modified. This leads us back to the functioning of the Sequential Drum. These working modes (transmission of the data when there is an external request or strike detection) allow a dialog, between the instruments and the device to be controlled. The microprocessor of the Radio Baton makes the implementation of this communication possible by the use of the MIDI protocol and of a program like MAX (Puckette and Zicarelli 1990) to communicate with the Radio Baton. Thus the control possibilities with instruments that can be programmed are almost unlimited. Numerous musical applications of the Radio Baton have already been published (see e.g. (Boie et al., 1989b), (Boulanger and Mathews, 1997),(Gershenfeld and Paradiso, 1997), (Jaffe and Schloss, 1994)) and most of them mainly exploit the drum-like structure of the device which is well adapted to the launching of sound events and to the conduction of their time evolution. We shall briefly describe how the Radio Baton can be diverted from its original aim and used to perform sound transformations with additive and physical synthesis techniques. A presentation of the Radio Baton piloting sound synthesis models was given by our research group at an international colloquium on new expressions related to music organized by GMEM (Groupe de Musique Expérimentale de Marseille) in Marseille (Kronland-Martinet et al.1999; Ystad, 1999). Intimate sound transformations using the Radio Baton were demonstrated. The instrument has also been used in a musical context with the goal of performing improvised sound transformations.

2.1 Radio Baton and Additive Resynthesis of Sounds

The amplitude modulation laws of the sound components together with their frequency modulation laws give the parameters defining an additive synthesis model. When connected to analysis techniques, additive synthesis methods give resynthesized sounds of high quality (Kronland-Martinet et al. 1997). However these models are difficult to manipulate because of the high number of parameters that intervenes. Mapping the Radio Baton to such a sound synthesis technique is closely linked to the compositional process since the choice of the parameters to be piloted is crucial to the way the instrument will respond and therefore determines the expressiveness offered by the instrument. We tried several mapping strategies, aiming at intimately control the sound by altering different parameters of the synthesis process. To pilot the sound of an additive synthesis model with the Radio Baton we chose to act on three parameters, i.e. the duration of the note, its fundamental frequency and its amplitude. In each case the manipulations can be done independently for each modulation law or globally on all the modulation laws. In our case the duration of the sound and the frequency manipulations are done in a global way. This corresponds to a simple acceleration or slowing down of a note when the duration is altered, and to a simple transposition when the frequency is altered. The amplitude modulation laws have been modified differently, giving the possibility of effectuating a filtering or equalization on the sound. In 2, the control possibilities of the Radio Baton are illustrated. The sound is generated when one of the sticks cuts a virtual plane the height of which is predefined by the user. The x-coordinate is related to the duration of the generated sound and the y-coordinate to the transposition factor. The second stick is used to control the note after it has been trigged (aftertouch) and uses the y coordinate to act on the frequency transposition (like for the first baton) and the z-coordinate to fix the slope of a straight equalization line. This slope is positive when the baton is over a predefined plane (0 point in 3) corresponding to an attenuation of low-frequency components and thus to a high-pass filtering. This is illustrated in 4. When the baton is below the zero point, the slope is negative, corresponding to a low-pass filtering. This allows a continuous modification of the brightness of the sound. The second stick could have controlled a third parameter corresponding to the x-coordinate, but since playing the Radio Baton was difficult, with the possibilities already described, this parameter was not used.

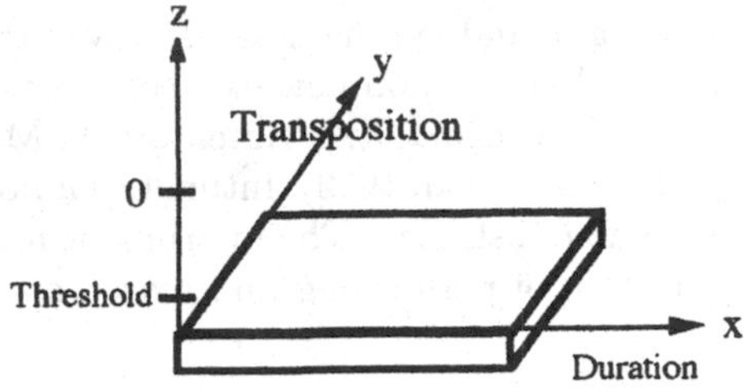

Fig. 2. Control of the additive synthesis instrument.

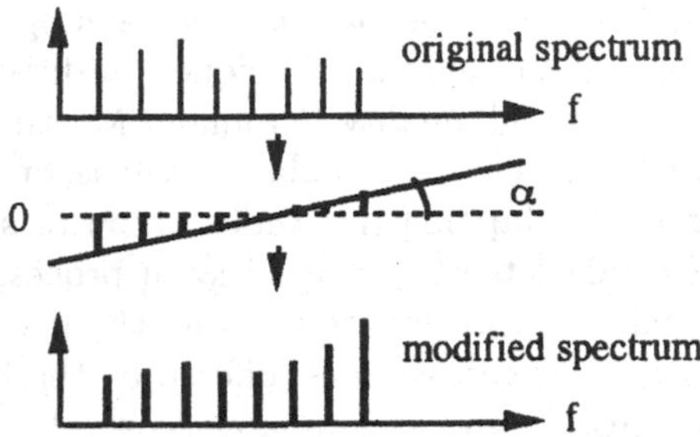

Fig. 3. Equalization of the spectrum over the zero point.

2.2 Radio Baton and Physical Modeling of Sounds

In a way similar to the description in the previous section, we have used the Radio Baton to pilot a physical synthesis model. This application tends to prove that physical models are well adapted to sound transformations by providing a small amount of meaningful parameters making the mapping easier. Physical models describe the sound generation system using physical considerations. An interesting method consists in simulating the way the acoustical waves propagate in the instrument by using a looped system with a delay line and a linear filter (5). Such a model is called the digital waveguide and it has been widely used for synthesis purposes (Smith 1992). The delay is proportional to the length of the resonator and thus to the fundamental frequency of the note played, while the filter takes into account the dissipation and the dispersion phenomena. Analysis-synthesis techniques adapted to the digital waveguide can be designed, allowing the resynthesis and the manipulation of a given natural sound (Kronland-Martinet et al. 1997; Ystad 2000). When the Radio Baton acts on a digital waveguide model the parameters to be modified have a physical significance. The parameters that are controlled by the sticks in this example only correspond to the delay and to the excitation of the model (the source), the filter being chosen once and for all. One of the batons acts on the choice of the excitation: each of the 4 corners of the receiver corresponds to a different excitation as shown in 6. In the middle of the plate, the source corresponds to a mixture of the four sources with weights depending on the distance from each corner. The second baton acts on the de-

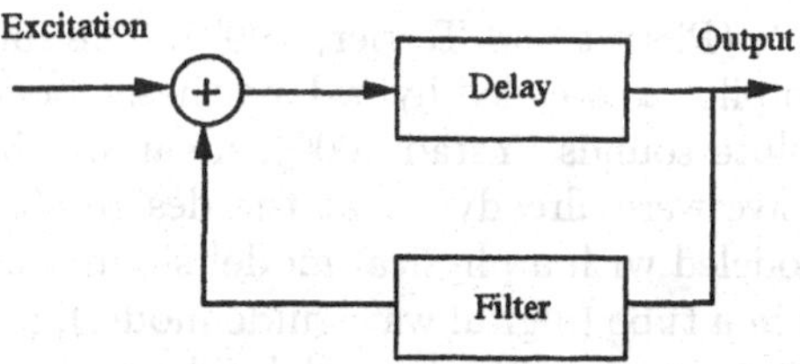

Fig. 4. Physical synthesis model.

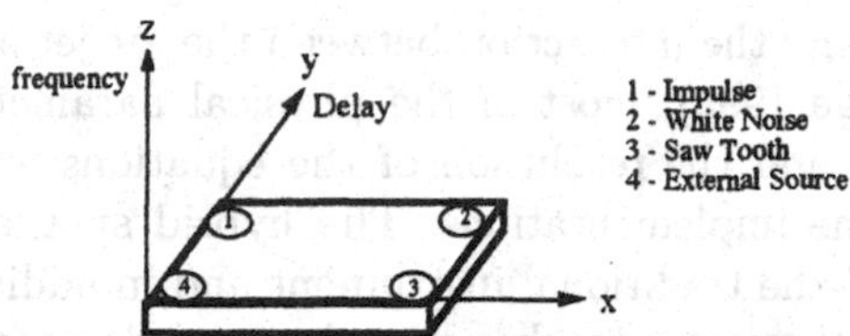

Fig. 5. Control of the instrument with a digital waveguide synthesis model.

lay of the resonator (thus on the frequency of the note played) given by the y-coordinate. In addition it acts on the frequency of the excitation signal when a saw tooth source is used. In this chapter we gave an example of how to divert an interface from its initial aim. The piloting of both the additive synthesis model and the physical model showed that the Radio Baton was an interesting tool for pedagogic purposes, but that it was difficult to seriously play with it because of the lack of absolute reference in the 3D space. Therefore this kind of approach is more adapted to improvisation purposes, without expecting to exactly reproduce a sequence. (Schloss, 1990), (Boulanger and Mathews, 1997) showed that the Radio Baton also is well adapted to control percussive sound models, while (Kronland-Martinet et al., 1999) successfully used it to generate new sounds that were not directly related to real instruments such as sounds produced by a FM synthesis model.

3 Extending the Possibilities of a Traditional Instrument: The Flute Case

The experience of the Radio Baton where new gestures had to be learned pushed us towards other types of interfaces, supposedly easier to use than the Radio Baton, namely interfaces adapted to specific gestures. Two different examples of such interfaces are described in this paper. The first one is made using a traditional flute connected to a computer by magnetic sensors detecting the finger position and a microphone at the embouchure level detecting the pressure variations (Ystad and Voinier 2001). An earlier attempt to extend the possibilities of a traditional flute was made at IRCAM (Pousset, 1992). However, this instrument was mainly made to act as a MIDI controller. Other attempts in designing meta-instruments have been made and the reader can for example refer to (Bromwich,

1997), (Jensen, 1996), (Pierrot and Terrier, 1997). The interface we developed has been designed to pilot a so-called hybrid synthesis model made to resynthesize and transform flute sounds (Ystad 2000), meaning that the mapping and the musical perspective were already set at the design stage. The resonator of the instrument is modeled with a physical model simulating the propagation of the acoustical waves in a tube (digital waveguide model), and the source is modeled using a non-linear signal synthesis model. The reason why a signal model has been used in this case is related to the fact that the physical phenomena observed at the embouchure of a flute are not fully understood. Even though some models describing the interaction between the air jet and the labium have been proposed (Verge 1995), most of the physical parameters intervening are difficult to measure, and the resolution of the equations is generally not compatible with real-time implementations. The hybrid synthesis model makes it possible to "imitate" the traditional instrument and in addition make very subtle modifications on the sound by changing the model's parameters. Hereby one can obtain physically meaningful sound transformations by acting on different parts of the model to simulate effects like exaggerated vibratos or change the properties of the source and/or the resonator. It is important to underline that the aim of such an interface is not to replace the traditional instrument, but to expand its possibilities. This augmented instrument will give musicians the possibility of making use of already acquired playing techniques. The flute has been equipped with Hall effect sensors detecting the distance to magnets connected to each keypad. The sensors have been placed on an aluminum rail fastened to the flute, while the magnets are fastened to the support rods where the keypads are fixed so that they approach the sensors when the keypads are closed as illustrated in 6 and 7. The state of the keypads is related to the frequency of the note played which means that the player's finger position will pilot the delay line of the resonator model and thus the frequency of the played note as shown in 8. The signal synthesis model which simulates the source of the instrument consists in two models - one which simulates the deterministic part of the source and one which simulates the stochastic part. The deterministic part of the model consists in a sine generator and a non-linear waveshaping function. The stochastic part of the model consists in a lowpass filtered white noise. These models are piloted by the pressure variations from the player's mouth which are measured by a microphone situated inside the flute at the cork position near the embouchure. In addition the sine generator of the deterministic part of the source model is piloted by the player's finger position. This system does not represent a complete way of characterizing the play, since for instance the angle at which the air jet hits the labium, or the position of the player's lips, are not taken into account. However, these important features influence the internal pressure which is measured by the microphone and which acts on the generated sound. When the state of the keypads does not change when the same note is played in different registers, the measurements of the pressure level at the embouchure allows the determination of the frequency of the note played. The speed at which the keypad is closed can be calculated by continuously measuring the distance

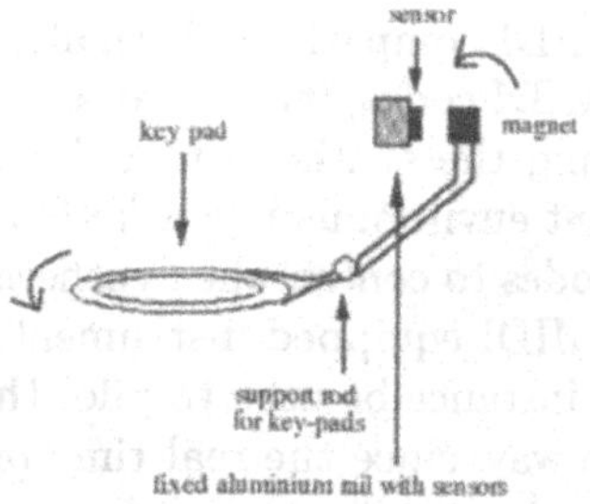

Fig. 6. Schematic illustration of the connection between the keypad and the magnet.

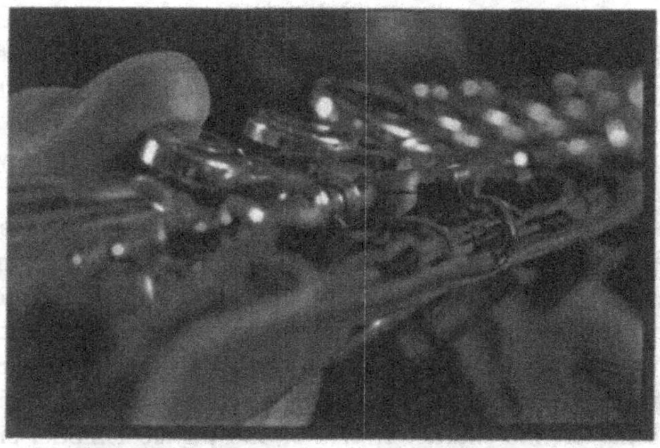

Fig. 7. Close view of a flute with magnets and sensors.

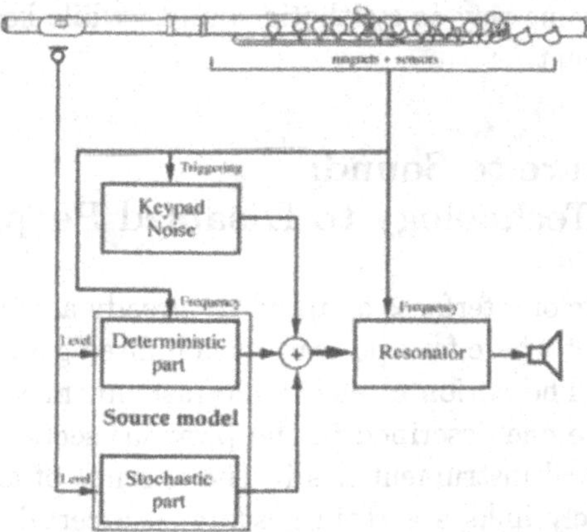

Fig. 8. Flute with synthesis model.

between the sensor and the magnet. Thus the keypadnoise, which represents an important part of the flute play, can be taken into account. Although the synthesis model is the most important part of the flute interface, it should be mentioned that it also is MIDI compatible. Actually, the data from the sensors are processed to generate MIDI codes, which are sent to the real-time processor. The MIDI flute interface and the synthesis model are both implemented using the MAX/MSP development environment (Puckette et al. 1990). Since the flute interface generates MIDI codes to control the synthesis model, one can use these codes to pilot any kind of MIDI equipped instrument. Thus, the pressure variations from the flute can for instance be used to pilot the tempo of an arpeggio on a MIDI piano. In the same way, since the real-time processor is also considered as a MIDI instrument, assigning arbitrary frequency values for a given key state can change the tune of the flute. At this point, the musical possibilities are only limited by the imagination. The parameters of the synthesis model can dramatically act on the sound itself. Since the digital flute model has been designed to respond to MIDI codes, one can act on the parameters of the model using MIDI controllers such as pedals, sliders, ...(Ystad and Voinier 2001). Cross synthesis effects can for instance be made this way by keeping the source model of the flute while replacing the tube model by a string model, or inversely, by injecting a voice signal or a source of another instrument into the resonator model of the flute. The last alternative would also allow the generation of the noises made by the flautist while playing. Actions on the non-linear source model make it possible to modify the spectral content of the sound so that clarinet-like sounds or brass effects can be obtained. The model we implemented also comprises a keypad noise generator, which can be altered by modifying the corresponding table. This keypad noise could for instance be replaced by any percussive sound. Finally, by acting on the characteristics of the physical model simulating the tube of the instrument, physically unrealizable instruments like a gigantic flute can be simulated. All these manipulations show the advantage of a hybrid sound model associated to a meta-instrument, which enables the modification of a natural sound in the same way as synthetic sounds, while keeping the spirit of the traditional instrument.

4 From Gesture to Sound: Adapting Technology to Disabled People

The second example of interfaces adapted to already acquired gestures is related to an artistic project where four persons with limited gestural faculties are piloting sound models. The notion of electronic instrument which is presented here is different from the one described in the previous section. The particularity of the electronic musical instrument lies in the absence of an acoustic mechanism system, which closely links a certain gesture to a certain sound event (the instrumental tone). A musician has to learn to play the instrument, by mastering a gestural repertoire that enables him to produce sounds with the instrument and to tackle its musical range. The problem of the electronic instrument depends to a certain extent on how much the different stages of monitoring the sound phenomena are left in the hands of the user/creator (gestural abilities, sound synthesis device). Between gesture and sound, one does not necessarily manipulate masses in movement (hammer, air column, string) but rather digital data. This concept has the effect of separating gesture and produced sound in their immediate (conventional) relationship, and new relationships between gesture, produced sound and musical creation therefore have to be invented. This project has been going on for about five years, with the participation of four persons with limited gesture faculties. The main features of this work (in continual development) lies in the fact that the technological realization is closely linked to the artistic problems and the heavy conditions of the handicap itself. We did not wish to give ourselves a particular goal as to the final form, but work patiently on the development and awareness, creating visual, sound and technological methods that would shape the rather indefinable realization (neither show nor installation) that we have called Un Bon Moment, A Sound and Vision Walk About. For the Australian Aboriginals, the notion of A Walk About denotes a short informal holiday period, far from work, when they can wander through the bush, visit relations, or go back to native life. Hence the project aims at offering the audience the idea of a walk about, where there is no longer a question of coherence with the idea of shape, but rather of an awareness of events that are about to happen in a space of time that belongs to the spectator himself, as he can go in and out of the performance area as he pleases. When designing interfaces for disabled persons it is important to be aware of the fact that it is the framework or the a priori concerning the results which creates the handicap rather than the person himself (if one considers him as such). For example, it is obvious that traditional musical instruments are all made with reference to specific gestural possibilities (movements and space between the arms, hands and fingers, the speed and precision of these movements, etc...). Even if the apprenticeship is long, an able-bodied person is able to play any traditional instrument. This is not the case for "someone with a handicap", who will neither necessarily adapt the characteristics of his body (size, space, force and precision) to the instruments, nor in the music that has been composed with or for these instruments. Therefore, the idea consisted in conceiving an instrument, which is neither a device that imposes a gestural form, nor a device adapted to specific morphological and motor abilities (as opposed to a traditional acoustic instrument), but rather a device able to "learn" the player's gestures. This means that the purpose of this project goes far beyond the question of "giving handicapped people access" to music (which in the long run relies on the instrumental capacities of an able-bodied person), and tackles the question of musicality which does not depend on virtuosity.

4.1 The Question of Time in Un Bon Moment

One of the initial observations made during this project, concerned a completely unusual relationship with time. It takes more time for handicapped than for able-bodied people to communicate (linked notably to difficulties in articulating), and therefore it takes more time to work together. The handicapped need some time between an idea and its expression, or a decision (to accomplish a gesture for example) and the moment when the decision really takes shape. Thus it prevents us from synchronizing the sound events in a fixed, predetermined order, and therefore from planning a musical score in its usual way. Finally, the notion of time is felt differently according to each person. It is related to the interior rhythm, underlined by music through beat, figure, gesture and movement. One can imagine that the notion of time for someone who can only move in a wheelchair since birth, or who can only talk very few words at a time in an extreme state of tension, is very different to that for an able-bodied person. Thus it seemed essential to us to not only take into account these particular relationships with time, but also to bring them to the foreground, both when creating the instrument and when realizing the performance.

4.2 The Gestures and the Music

Obviously, the player's gesture is no longer just the one of a performer involved in the mere production of a piece in which the organization of its constitutive elements has been preconceived. This gesture becomes a significant element in a sound production no longer thought in terms of organised objects but rather in terms of a sensitive sound experience set in time and space. The gesture, which consists in producing a sound, can therefore be considered as the origin of the shape which is afterwards gradually elaborated. A lot of work related to gestures and music has recently been published and we refer the reader for example to (Camurri et al., 1998), (Coutaz and Crowley, 1995), (Azarbayejani et al., 1996), (Sawada et al., 1997) for the state of the art in this field. In Un Bon Moment, our main interest is precisely to use the gesture which produces and controls the audio-visual events as a source of creation. The gestural model is then free from any conventions (and particularly from musical ones). Again, the main purpose in Un Bon Moment is not to integrate these unexpected gestures into the framework of preexisting musical productions, but to use them in the implementation of specific musical means, as factors of artistic invention and reflection. In this context, impossibilities of determining time and synchronizing events become elements which could involve situations of sensitive creation.

4.3 The Actual State of the Project

So far, two public performances of "Un Bon Moment" over a period of about one month have been carried out. Four performers, having few and particular gestural possibilities were running moving machines (called "Mobil Home"), each one being a sculpture based on a wheelchair, equipped with instrumental devices. One of them was intended to record, transform and project images, and the three others were more specifically intended for playing sounds. The three instrumental sound devices were conceived with the same idea: a gesture sensor, a module which transforms information given by the sensor into MIDI data, a sound generating module and a video projector which projects the whole or a part of the computer screen. The three gesture sensors are:

- A Headmouse System (Origin Instruments), controlling the mouse pointer on the computer screen, by means of an optical sensor following the movements of a disc stuck on the operator's forehead.
- A lever, large-sized (and large range), as a joystick.
- A breath and lip-pressure sensor, which is part of a Yamaha WX7 wind controller.

These three gesture sensors communicate with the MAX software, either directly in the case of the WX7, or through an I-Cube (Infusion Systems) in the case of the lever, or with a MAX object (the PAD object) especially created for this work. For the moment, the sound generating modules for the three instruments, are supplied by the audio section of the MAX (MSP) software. The computer screen is projected either on a surface which has been integrated into the mobile instrumental device, or on an outside one. It enables the visualization of texts which are heard and treated in a sonorous way, of images which have been transformed by MIDI data picked up by the gesture sensor (Imagine Software), and of the redesigned mouse pointer which will be described more precesely in section 4.4. An important part of the device is the so called PAD object which is an external rectangular graphical object. Like all the MAX objects, it can be moved on the patch while it keeps its intrinsic characteristics. It is resizable by a simple mouse click and slide. The PAD object analyzes the movements and actions of the pointer located in the screen zone that it covers, and sends back the mouse's pointer position relative to the lower left hand corner (an offset and a multiplying factor can also be given), as well as the speed and acceleration of the pointer's movement. In addition a message is given if the pointer enters the object or leaves it, if a click occurs when the pointer is inside the object or if the mouse button is held down when the pointer is inside the object and moves (click and drag).

4.4 Rafika Sahli-Kaddour's Instrument

Rafika Sahli-Kaddour is one of the four performers. She has the ability and practice to drive a computer mouse cursor with her head, using a HEADMOUSE system. This gesture sensor enables her by the use of the PAD object to act upon three instrumental devices working differently. The first device projects the computer screen on a surface or a space where visual indicators (images, objects, etc.) have been set up: Each of them corresponds to invisible PAD objects. In this way, Rafika can direct the mouse pointer (redesigned as a red circle) towards real objects in space, move it on to the objects and produce and control the sound events according to three visual indication points. The projected image can thus be considered as a "non-linear score". The second device is fairly similar to the first, apart from the fact that a computer screen image is projected (it could be a film), and the objects that allow the control of sound events can correspond to certain zones of the image (or the film). The third device does not need a screen projection. It makes use of the speed information and the horizontal or vertical movement of the pointer given by the PAD object. The player controls sound transmissions by the head movements and the speed of these movements. In this chapter we have described how devices able to 'learn' the player's gestures can be made. Such interfaces allow new musical gestures outside the realism of conventional musical attitudes as well as the creation of other forms of musical script linked to imprecision and approximation. They also show the need of a reflection on musical choices simultaneously with the creation of instrumental devices confirming the fact that through the concept of an instrument, one has already started to compose the music with it. Finally, beyond the opposition between acousmatics and live instrumental music, the playing of an electronic musical instrument pushes us to put forward a more fundamental aspect - more innovative - which lies in the fact that the composer-musician (and/or improviser) directly can intervene on the sounds (sound phenomena), precisely at the moment when they develop, last, and establish themselves. In this way an idea or a musical form can be constructed and developed on the basis of the sensation of duration, and of the modes of articulation, in contrast to a structure of logic and rhetoric. These playing techniques closely link sound matter and musical composition together, (both are generated at the same time), and evidently question the relationships between organization - the implementation of rules- and sound material in a new way. Hence, this way of creating music implies that the work is considered as a living moment, whereas the music created for traditional instruments is considered as a constructed object.

5 The "Particle Motion": The Cosmophone

So far the interfaces that have been described are to be piloted by people. In the last example we describe a device which is piloted by natural phenomena. Many composers have used random phenomena to compose music, from Mozart with his "Musical Dices Game" to John Cage with his piece "Reunion", where the player moves on a specially equipped chessboard to trigger sounds. Natural random-like phenomena have also been used, as in the piece entitled "The Earth's Magnetic Field" by Charles Dodge for which a computer translated fluctuations in the magnetic field of the Earth into music. Many other concepts for sound generation from such a source of random events can be (and are actually) explored. The interface we have developed and which is called the "Cosmophone" was initially made to make people aware of the cosmic rays that collide with the earth's atmosphere and fall on the surface of the earth.

5.1 The Cosmic Rays

Interstellar space is filled with a permanent flux of high-energy elementary particles called "cosmic rays". These particles have been created by violent phenomena somewhere in our galaxy, for example when an old massive star explodes into a supernova. The particles then stay confined in the galaxy for millions of years by the galactic magnetic fields before reaching our planet. When colliding against the earth's atmosphere, cosmic rays create showers of secondary particles. Though partly absorbed by the atmosphere, these showers induce a large variety of phenomena, which are measurable at the sea level. The main phenomenon is a flux of muons, a kind of heavy electron absent from usual matter because of its short lifetime. Muons are produced at a large rate in cosmic showers. Thanks to their outstanding penetrating properties, they are able to reach the ground. At the sea level, their flux is about hundred muons per second per square meter. Highly energetic cosmic rays produce bunches of muons, or multi-muons, having the same direction and a few meters apart from each other. The number of muons within a bunch is a function of the energy of the primary cosmic ray. Within an area of a hundred square meters, the rate of multi-muons bunches ranges from one per second (bunches of two or three muons) to one per minute (bunches of ten muons or more). Muon interaction within the matter is another phenomena. When muons pass close to atomic nuclei, electromagnetic showers composed of electron-antielectron pairs are created. This phenomenon can be observed for example inside buildings with metallic structures, at a rate of about one per minute per ten square meters.

5.2 The Concept of Cosmophone

Human beings are insensitive to particles passing through their body. The Cosmophone is a device designed to make the flux and properties of cosmic rays directly perceptible within a three dimensional space. This is done by coupling a set of elementary particle detectors to an array of loudspeakers by a real time data acquisition system and a real time sound synthesis system. In that way, information received from the detectors triggers the emission of sounds, depending on the parameters of the detected particles. These parameters and the rate of occurrence of the different cosmic phenomena allow a large variety of sound effects to be produced. Because of the fluctuations in the occurrence of the phenomena, the set of detectors can be seen as a set of random number generators.

5.3 Sound Generation and Spatialization in the Cosmophone

According to the concept explained above, the synthesis system has to generate sounds when triggered by the particle detection system. To simulate a particle rain, in which listeners are immersed, we have grouped the loudspeakers in two arrays; one above the listeners (placed above a ceiling), and the other one below them (placed under a specially built floor). The arrays of loudspeakers are disposed so that the ears of the listeners (who are supposed to be standing up and moving inside the installation), are located approximately at an equal distance from the two groups. Both ceiling and floor are acoustically transparent, but the speakers are invisible to the listeners. A particle detector is placed close to each loudspeaker. When a particle first passes through a detector of the top group, then through a detector of the bottom group, a sound event is triggered. This sound event consists in a sound moving from the ceiling to the floor, "materializing" the trajectory of the particle. Because of the morphology of the human ears, our hearing system can accurately localize moving sources in a horizontal plane, but is far less accurate in the vertical plane. Initial experiments have shown us that the use of a panpot to distribute the signal energy between two loudspeakers was not sufficient to create the illusion of a vertically moving sound source. To improve the illusion of a vertical movement, we have used the Doppler effect, which is perceivable when there is a relative movement between an acoustic source and a listener. It then leads to a modification of the pitch of the perceived sound during time, as well as a modification of its amplitude. This effect is very common in every day life, and our hearing system is used to recognizing it. Chowning (Chowning-71) has shown that this effect is essential for the realism of moving source simulation. Using a Doppler effect simulation along with the energy panpot between the ceiling and the floor speakers greatly improve the illusion of a vertical movement of the sound source. But the departure and arrival points of the moving source in the space remain rather imprecise for listeners. To improve the localization, we have then added two short sounds as starting and ending cues. The first cue is emitted from the high loudspeaker at the beginning of the sound event; the second comes from the low loudspeaker, at the end of the event. Because they are chosen to be very precisely localizable, these two cues greatly improve the illusion, giving the impression of a sound crossing the ceiling, then hitting the floor. A particular Cosmophone installation has been made for the Cité des Sciences et de l'Industrie in Paris, and is part of an exposition area on particle physics called the "Thétre des Muons" (9). For this installation, two arrays of twelve speakers and detectors are in two concentric circles; the inner one comprises four speakers and detectors, the outer one the eight others. The outer circle is about five meters in diameter, allowing several listeners to stand in the installation. This installation is open to the public in the Cité des Sciences et de l'Industrie in Paris. A prototype composed of two arrays of four detectors and speakers is actually installed at L.M.A. in Marseille.

Fig. 9. Picture of the Cosmophone installation in the Cité des Sciences et de l'Industrie in Paris.

5.4 The Particle Detection System

Each particle detector is composed of a slat of plastic scintillator associated to a photomultiplicator. A particle passing through the scintillator triggers the emission of a photon inside the scintillator. The photon is then guided into the scintillator until it reaches the photomultiplicator. An electrical impulse is then generated. This kind of detector is very sensitive to ambient light and radioactivity. We can easily protect it from ambient light, but to be sure to detect only particles generated by cosmic rays, a coincidence system is used. Such particles are first passing through a detector located in the ceiling, then through a detector located under the floor. Knowing that these particles travel at light speed, and knowing the distance between high and low detectors, the two events occurring in a given time (a few nanoseconds) is a signature of a cosmic induced particle. The coincidence triggers the readout of all the detector signals, which allows identifying the speakers in which the sound is to be produced. These signals are transmitted to a PC computer and processed to identify the kind of event detected, single particle, or bunch(es) of multi-muons. Depending on the energy of the cosmic rays and on their occurrence, it is possible to observe during a single event a combination of single muons and bunches of multi-muons. The program makes use of its knowledge of the geometrical arrangement of the detectors to recognize which phenomena were detected. Depending on the location of the installation, it is possible to observe a high rate of particle, which would generate a high rate of sound events and make the listeners rather confused. We can therefore decide to ignore some of these events to avoid overloading the listeners auditory systems. Finally, the information is sent to the sound synthesis system through a MIDI interface, using a custom protocol.

5.5 The Sound Synthesis System

The sound synthesis system is built on an Apple Macintosh computer, running the MAX/MSP real time synthesis software. The computer is equipped with a MOTU (Mark Of The Unicorn) audio interface which is able to output simultaneously twenty-four channels of audio signals, which are sent to the amplifiers and loudspeakers. The synthesis program is composed of a set of modules or instruments; each of them is able to generate sounds associated with one event. The more powerful the computer, the more instruments can be run at the same time, and the more events can be played simultaneously. As mentioned above, different phenomena are detectable. In particular three cases were distinguished: a single muon reaching a pair of detectors (high then low); a "small bunch", where more than one pair of detectors hit simultaneously, but less than four, and a "big bunch", when at least four pairs are hit. We decided that the three cases would be illustrated by different sound sequences. When using a lot of detectors, the three events may occur simultaneously, which means that one instrument should be able to generate several sound sequences at the same time. A dynamic router has been implemented to allocate one sound event to one free instrument by passing the incoming MIDI data on to it. Therefore the instrument becomes busy during the sound generation process then returns to a free state. An instrument consists in a scheduler that triggers the three successive sounds: the cue in the ceiling speaker, followed by the sound moving from the ceiling to the floor, and finally by the ending cue in the floor speaker. Depending on the event's content, ups to three sequences are played simultaneously. The scheduler manages the appropriate timing of the sequences by sending start messages to the appropriate sound players. Two kinds of sound players are used: the cue sound players and the moving sound player. The cue sound player is very simple: it reads a memory stored sound sample. The moving sound player is a little bit more complicated, as it has to apply the Doppler frequency shift and amplitude modification. It makes use of a variable speed sound sample player and pre-computed tables for frequency and amplitude modifications. The final part of the instrument is an audio signal router that feeds the appropriate output channel, according to the information received for the event.

5.6 The Musical Potential

The final sound quality of this installation mainly depends on the nature of the sounds associated with the detected events. Since real particles travel at light speed and do not generate audible sounds, the real phenomenon could not be reproduced. Sounds therefore had to be created to give a mental representation to the listener. The main difficulty consists in creating appropriate moving and cue sounds, to call to mind the image of a particle rain. A lot of sounds were created by different synthesis techniques, and about ten moving sounds and ten cue sounds were selected and submitted to a panel of listeners for final judgement. We have tested many associations of moving and cue sounds to obtain the final result. The synthesis system allows fine-tuning of the Doppler effect parameters, which has appeared to be very useful for the final "tuning" of the sound impression. To help listeners concentrate on their auditive impressions when entering this installation, we have decided to diffuse some continuous background music through all the speakers. The installation has appeared to be a helpful tool for music spatialization. One of the authors has composed an original music score to be played in the Cosmophone, taking advantage of these possibilities.

6 Conclusion

The four developments that are presented in this article lead, by different approaches, to a conclusion where we discuss the musical intention behind computer systems used as musical instruments. The creation of electronic instruments, which from our point of view is a part of the musical composition, inevitably puts forward questions about traditional relations between movements and sound production, thoughts and musical realizations. The main difference between electronic and acoustic instruments seems to be the relation between the gestures and the sounds. What acoustic instruments are concerned, there is a one-to-one relation between these two aspects since a particular gesture generates a particular sound and since variations produced by muscular energy are closely related to variations in timbre, intensity and pitch. When it comes to electronic instruments these relations no longer exist, and everything has to be invented. This leads to a new way of considering the interface as an element of the musical composition. Even though this article aims at describing different approaches of interface design, we always kept the relationship between music and events in mind. In the first example we used an already existing interface, the Radio Baton, to pilot different synthesis models. At an early stage we realized that this interface was restrained due to the fact that it, in the first place, was intended to conduct musical sequences or to be used as a drum-like instrument. Diverting this interface from its primary goal showed us the danger of rapidly falling into a "spectacular" use rather than a musical one. Nevertheless, its use for sound manipulation in an improvising context was found interesting. The second approach related to interfaces adapted to specific gestures dealt with two different kinds of interfaces. In the first example we constructed an augmented flute to give already skilled musicians access to new technologies. These interfaces mainly are of interest when they are related to synthesis models which correspond to the behavior of the instrument. This makes such controllers well adapted to physical synthesis modeling which naturally increases the musical possibilities of the instrument. In the second example of adapted interfaces we showed how interfaces can be adapted to disabled persons. This is a good illustration of how both the gesture and the musical goal should be taken into account when constructing interfaces, since a different handicap imposes a different interface. The last approach concerns interfaces piloted by natural phenomena. As an example we describe the Cosmophone which represents an extreme case where no gestures are needed, and where natural phenomena generate a three dimensional sensation of the trajectories of cosmic rays. In this case a specific set of detectors associated to an adequate mapping directly creates this objective. These examples show that the musical dimension has to be part of the conception of an interface, and led us to the conclusion that: even though new technologies provide numerous devices which can be used to pilot sound processors, a genuine musical interface should go past the technical stage to integrate the creative thought.

Acknowledgments

The Cosmophone was designed from an original idea of C. Vallée (Centre de Physique des Particules de Marseille) who built the particle detector together with D. Calvet.

References

1. Azarbayejani, A. Wren, C. Pentland, A. 1996. Real-time 3-D tracking of the human body, in Proceedings of the IMAGE'COM96, May 1996.
2. Battier M. 1999. L'approche gestuelle dans l'histoire de la lutherie électronique. Etude d'un cas : le theremin. in R. de Vivo and H. Genevois (eds) Les nouveaux gestes de la musique, Editions Parenthèses, Collection Eupalinos, ISBN 2-86364-616-8.

3. Boie, R. Mathews, M. Schloss, A. The radio drum as a synthesizer controller, in Proc. Int. Computer Music Conf. (ICMC'89), pp. 42-45, 1989b.
4. Boie, R.A. Ruedisueli, L.W. and Wagner, E.R. 1989a. Gesture Sensing via Capacitive Moments. Work Project N 311401 AT& T Bell Laboratories.
5. Boulanger, R. Mathews, M. 1997. The mathews' radio baton and improvisation modes, in Proc. Int. Computer Music Conf. (ICMC'97), pp. 395-398.
6. Bromwich, M.-A. 1997. The metabone: An interactive sensory control mechanism for virtuoso trombone. in Proc. Int. Computer Music Conf. (ICMC'97), pp. 473-475.
7. Camurri, A. Ricchetti, M. Di Stefano, M. Stroscio, A. 1998. EyesWeb - toward gesture and affect recognition in dance/music interactive systems, in Proc. Colloquio di Informatica Musicale CIM'98, AIMI.
8. Chowning, J. 1971. The Simulation of Moving Sound Sources. Journal of the Audio Engineering Society 19:1.
9. Coutaz, J. Crowley, J. 1995. Interpreting human gesture with computer vision, in Proc. Conf. on Human Factors in Computing Systems (CHI'95), 1995.
10. Gershenfeld, N. Paradiso, J. 1997. Musical applications of electric field sensing, Computer Music J., vol. 21, no. 2, pp. 69-89, 1997.
11. Jaffe, D.A., Schloss, W.A 1994. A Virtual Piano Concerto in Proc. Int. Computer Music Conf. (ICMC'94), 1994.
12. Jensen, K. 1996. The control mechanism of the violin. in Proceedings of the Nordic Acoustic Meeting, (Helsinki, Finland), pp. 373-378, 1996.
13. Kanamori, T. Katayose, H. Simura, S. Inokuchi, S. 1993. Gesture sensor in virtual performer, in Proc. Int. Computer Music Conf. (ICMC'93), pp. 127-129, 1993.
14. Katayose, H. Kanamori,T. Simura, S. , and Inokuchi, S. 1994. Demonstration of Gesture Sensors for the Shakuhachi. In Proceedings of the 1994 International Computer Music Conference. San Francisco, International Computer Music Association, pp. 196-199.
15. Keane D. and Gross, P. 1989. The MIDI baton. in Proc. Int. Computer Music Conf. (ICMC'89), pp. 151-154, 1989.
16. Kronland-Martinet R., Guillemain Ph., Ystad S. 1997 "Modelling of Natural Sounds Using Time-Frequency and Wavelet Representations" Organised Ssound, Vol.2 n3, pp.179-191, Cambridge University Press.
17. Kronland-Martinet, R. Voinier, T. Guillemain, P. 1999. Agir sur le son avec la baguette radio in R. de Vivo and H. Genevois (eds) Les nouveaux gestes de la musique, Editions Parenthèses, Collection Eupalinos, ISBN 2-86364-616-8.
18. Laubier, S. de 1999. Le Méta-Instrument a-t-il un son? Emergence de lois ou de constantes dans le développement d'instruments virtuels. In H. Genevois and R. de Vivo, (eds) Les nouveaux gestes de la musique. Marseille: Editions Parenthèses, pp. 151-156.
19. Machover, T. 1992. Hyperinstruments - a Composer's Approach to the Evolution of Intelligent Musical Instruments. In L. Jacobson, ed. Cyberarts: Exploring Arts and Technology. San Francisco: MillerFreeman Inc., pp. 67-76.
20. Mathews, M.V., Schloss, W.A. 1989. The Radio Drum as a Synthetizer Controler in Proc. Int. Computer Music Conf. (ICMC'89), 1989.
21. Mathews, M.V. 1991a. The Conductor Program and Mechanical Baton. in M. Mathew and J. Pierce (eds) Current Directions in Computer Music Research. Cambridge, MA, MIT Press, 1991.
22. Mathews, M.V. 1991b. The Radio Baton and Conductor Program or: Pitch, the Most Important and Least Expressive Part of Music. Computer Music Journal 15:4.

23. Mathews, M.V. 1997. Exécution en direct à l'ge de l'ordinateur. in H. Dufourt and J.M. Fauquet (eds) La musique depuis 1945 - matériau, esthétique, perception. Mardaga, Brussels, 1997.
24. Mathews, M.V. Abbot C. 1980. The Sequential Drum. Computer Music Journal 4:4.
25. Moog, R. 1987. Position and Force Sensors and their Application to Keyboards and Related Controllers. In Proceedings of the AES 5th International Conference. New York, NY: Audio Engineering Society, pp. 179-181.
26. Moog, R., and T. Rea. 1990. Evolution of the Keyboard Interface: The Bosendorfer 290SE Recording Piano and the Moog Multiply-Touch-Sensitive Keyboards. Computer Music Journal, 14(2):52-60.
27. Moore, F. R. 1987. The disfunctions of MIDI. in Proc. Int. Computer Music Conf. (ICMC'87), pp. 256-262, 1987
28. Mulder, A. 1995. The I-Cube system: moving towards sensor technology for artists. in Proceedings of the ISEA, 1995.
29. Pierrot P., Terrier A. 1997. Le violon MIDI. tech. rep., IRCAM, 1997.
30. Pousset, D. 1992. La flte-midi, l'histoire et quelques applications. Mémoire de Matrise, 1992. Université Paris-Sorbonne.
31. Puckette, M. Zicarelli, D. 1990. Max - an Interactive Graphic Programming Environment. Opcode Systems.
32. Rovan, J., Wanderley, M. Dubnov, S. Depalle, P. 1997. Instrumental Gestural Mapping Strategies as Expressivity Determinants in Computer Music Performance. Kansei, The Technology of Emotion. Proceedings of the AIMI International Workshop, A. Camurri, ed. Genoa: Associazione di Informatica Musicale Italiana, October 3-4, 1997, pp. 68-73.
33. Sawada, H. Onoe, N. Hashimoto, S. 1997. Sounds in Hands - a Sound Modifier using Datagloves and Twiddle Interface. In Proceedings of the 1997 International Computer Music Conference. San Francisco, International Computer Music Association, pp. 309-312.
34. Schloss, A. 1990. Recent Advances in the Coupling of the Langage MAX with the Mathews/Boie Radio Drum. Proceedings of the International Computer Music Conference 1990.
35. Smith J.O. 1992. Physical Modeling Using Digital Waveguides. Computer Music Journal 16:4.
36. Smith, J. R. 1996. Field mice: Extracting hand geometry from electric field measurements, IBM Systems Journal, vol. 35, no. 3/4, pp. 587-608, 1996.
37. Snell, J. 1983. Sensors for Playing Computer Music with Expression. In Proceedings of the 1983 International Computer Music Conference. San Francisco, International Computer Music Association.
38. Verge, M.P. 1995. Aeroacoustics of Confined Jets with Applications to the Physical Modeling of Recorder-like Instruments. PhD thesis, Eindhoven University.
39. Vergez, C. 2000. Trompette et trompettiste: un système dynamique non linéaire analysé, modelisé et simulé dans un contexte musical. Ph.D. thesis, Université de Paris VI.
40. Wanderley, M. M. Battier, M. eds. 2000. Trends in Gestural Control of Music. Paris: Ircam - Centre Pompidou.
41. Wanderley, M. M. Viollet, J.-P. Isart, F. Rodet, X. 2000. On the Choice of Transducer Technologies for Specific Musical Functions. In Proceedings of the 2000 International Computer Music Conference. San Francisco, CA: International Computer Music Association, pp. 244-247.

42. Wright, M. Wessel D. Freed, A. 1997. New musical control structures from standard gestural controllers, in Proc. Int. Computer Music Conf. (ICMC'97), pp. 387-390.
43. Ystad, S. 1999. De la facture informatique au jeu instrumental. in R. de Vivo and H. Genevois (eds) Les nouveaux gestes de la musique, Editions Parenthèses, Collection Eupalinos, ISBN 2-86364-616-8.
44. Ystad, S. Voinier, T. 2000. A Virtually-Real Flute. Computer Music Journal (MIT Press), 25:2, pp 13-24, Summer 2001.
45. Ystad, S. 2000. Sound Modeling Applied to Flute Sounds, Journal of Audio Engineering Society, Vol. 48, No. 9, pp. 810-825, september 2000.
46. Yunik, M. Borys M. Swift G.W. 1985. 3A Digital Flute. Computer Music Journal 9(2): 49-52.

Author Index

Lecture Notes in Computer Science

For information about Vols. 1–2827

please contact your bookseller or Springer-Verlag

Vol. 2923: V. Lifschitz, I. Niemelä (Eds.), Logic Programming and Nonmonotonic Reasoning. IX, 365 pages. 2004. (Subseries LNAI).

Vol. 2916: C. Palamidessi (Eds.), Logic Programming. Proceedings, 2003. XII, 520 pages. 2003.

Vol. 2914: P.K. Pandya, J. Radhakrishnan (Eds.), FST TCS 2003: Foundations of Software Technology and Theoretical Computer Science. Proceedings, 2003. XIII, 446 pages. 2003.

Vol. 2913: T.M. Pinkston, V.K. Prasanna (Eds.), High Performance Computing - HiPC 2003. Proceedings, 2003. XX, 512 pages. 2003. (Subseries LNAI).

Vol. 2911: T.M.T. Sembok, H.B. Zaman, H. Chen, S.R. Urs, S.H. Myaeng (Eds.), Digital Libraries: Technology and Management of Indigenous Knowledge for Global Access. Proceedings, 2003. XX, 703 pages. 2003.

Vol. 2910: M.E. Orlowska, S. Weerawarana, M.M.P. Papazoglou, J. Yang (Eds.), Service-Oriented Computing - ICSOC 2003. Proceedings, 2003. XIV, 576 pages. 2003.

Vol. 2908: K. Chae, M. Yung (Eds.), Information Security Applications. XII, 506 pages. 2004.

Vol. 2906: T. Ibaraki, N. Katoh, H. Ono (Eds.), Algorithms and Computation. Proceedings, 2003. XVII, 748 pages. 2003.

Vol. 2905: A. Sanfeliu, J. Ruiz-Shulcloper (Eds.), Progress in Pattern Recognition, Speech and Image Analysis. XVII, 693 pages. 2003.

Vol. 2904: T. Johansson, S. Maitra (Eds.), Progress in Cryptology - INDOCRYPT 2003. Proceedings, 2003. XI, 431 pages. 2003.

Vol. 2903: T.D. Gedeon, L.C.C. Fung (Eds.), AI 2003: Advances in Artificial Intelligence. Proceedings, 2003. XVI, 1075 pages. 2003. (Subseries LNAI).

Vol. 2902: F.M. Pires, S.P. Abreu (Eds.), Progress in Artificial Intelligence. Proceedings, 2003. XV, 504 pages. 2003. (Subseries LNAI).

Vol. 2901: F. Bry, N. Henze, J. Ma luszyński (Eds.), Principles and Practice of Semantic Web Reasoning. Proceedings, 2003. X, 209 pages. 2003.

Vol. 2900: M. Bidoit, P.D. Mosses (Eds.), Casl User Manual. XIII, 240 pages. 2004.

Vol. 2899: G. Ventre, R. Canonico (Eds.), Interactive Multimedia on Next Generation Networks. Proceedings, 2003. XIV, 420 pages. 2003.

Vol. 2898: K.G. Paterson (Eds.), Cryptography and Coding. Proceedings, 2003. IX, 385 pages. 2003.

Vol. 2897: O. Balet, G. Subsol, P. Torguet (Eds.), Virtual Storytelling. Proceedings, 2003. XI, 240 pages. 2003.

Vol. 2896: V.A. Saraswat (Eds.), Advances in Computing Science – ASIAN 2003. Proceedings, 2003. VIII, 305 pages. 2003.

Vol. 2895: A. Ohori (Eds.), Programming Languages and Systems. Proceedings, 2003. XIII, 427 pages. 2003.

Vol. 2894: C.S. Laih (Eds.), Advances in Cryptology - ASIACRYPT 2003. Proceedings, 2003. XIII, 543 pages. 2003.

Vol. 2893: J.-B. Stefani, I. Demeure, D. Hagimont (Eds.), Distributed Applications and Interoperable Systems. Proceedings, 2003. XIII, 311 pages. 2003.

Vol. 2892: F. Dau, The Logic System of Concept Graphs with Negation. XI, 213 pages. 2003. (Subseries LNAI).

Vol. 2891: J. Lee, M. Barley (Eds.), Intelligent Agents and Multi-Agent Systems. Proceedings, 2003. X, 215 pages. 2003. (Subseries LNAI).

Vol. 2890: M. Broy, A.V. Zamulin (Eds.), Perspectives of System Informatics. XV, 572 pages. 2003.

Vol. 2889: R. Meersman, Z. Tari (Eds.), On The Move to Meaningful Internet Systems 2003: OTM 2003 Workshops. Proceedings, 2003. XIX, 1071 pages. 2003.

Vol. 2888: R. Meersman, Z. Tari, D.C. Schmidt (Eds.), On The Move to Meaningful Internet Systems 2003: CoopIS, DOA, and ODBASE. Proceedings, 2003. XXI, 1546 pages. 2003.

Vol. 2887: T. Johansson (Eds.), Fast Software Encryption. IX, 397 pages. 2003.

Vol. 2886: I. Nyström, G. Sanniti di Baja, S. Svensson (Eds.), Discrete Geometry for Computer Imagery. Proceedings, 2003. XII, 556 pages. 2003.

Vol. 2885: J.S. Dong, J. Woodcock (Eds.), Formal Methods and Software Engineering. Proceedings, 2003. XI, 683 pages. 2003.

Vol. 2884: E. Najm, U. Nestmann, P. Stevens (Eds.), Formal Methods for Open Object-Based Distributed Systems. Proceedings, 2003. X, 293 pages. 2003.

Vol. 2883: J. Schaeffer, M. Müller, Y. Björnsson (Eds.), Computers and Games. XI, 431 pages. 2003.

Vol. 2882: D. Veit, Matchmaking in Electronic Markets. XV, 180 pages. 2003. (Subseries LNAI).

Vol. 2881: E. Horlait, T. Magedanz, R.H. Glitho (Eds.), Mobile Agents for Telecommunication Applications. Proceedings, 2003. IX, 297 pages. 2003.

Vol. 2880: H.L. Bodlaender (Eds.), Graph-Theoretic Concepts in Computer Science. XI, 386 pages. 2003.

Vol. 2879: R.E. Ellis, T.M. Peters (Eds.), Medical Image Computing and Computer-Assisted Intervention - MICCAI 2003. Proceedings, 2003. XXXIV, 1003 pages. 2003.

Vol. 2878: R.E. Ellis, T.M. Peters (Eds.), Medical Image Computing and Computer-Assisted Intervention - MICCAI 2003. Proceedings, 2003. XXXIII, 819 pages. 2003.

Vol. 2877: T. Böhme, G. Heyer, H. Unger (Eds.), Innovative Internet Community Systems. VIII, 263 pages. 2003.

Vol. 2876: M. Schroeder, G. Wagner (Eds.), Rules and Rule Markup Languages for the Semantic Web. Proceedings, 2003. VII, 173 pages. 2003.

Vol. 2875: E. Aarts, R. Collier, E.v. Loenen, B.d. Ruyter (Eds.), Ambient Intelligence. Proceedings, 2003. XI, 432 pages. 2003.

Vol. 2874: C. Priami (Eds.), Global Computing. XIX, 255 pages. 2003.

Vol. 2871: N. Zhong, Z.W. Raś, S. Tsumoto, E. Suzuki (Eds.), Foundations of Intelligent Systems. Proceedings, 2003. XV, 697 pages. 2003. (Subseries LNAI).

Vol. 2870: D. Fensel, K.P. Sycara, J. Mylopoulos (Eds.), The Semantic Web - ISWC 2003. Proceedings, 2003. XV, 931 pages. 2003.

Vol. 2869: A. Yazici, C. Şener (Eds.), Computer and Information Sciences - ISCIS 2003. Proceedings, 2003. XIX, 1110 pages. 2003.

Vol. 2868: P. Perner, R. Brause, H.-G. Holzhütter (Eds.), Medical Data Analysis. Proceedings, 2003. VIII, 127 pages. 2003.

Vol. 2866: J. Akiyama, M. Kano (Eds.), Discrete and Computational Geometry. VIII, 285 pages. 2003.

Vol. 2865: S. Pierre, M. Barbeau, E. Kranakis (Eds.), Ad-Hoc, Mobile, and Wireless Networks. Proceedings, 2003. X, 293 pages. 2003.

Vol. 2864: A.K. Dey, A. Schmidt, J.F. McCarthy (Eds.), UbiComp 2003: Ubiquitous Computing. Proceedings, 2003. XVII, 368 pages. 2003.

Vol. 2863: P. Stevens, J. Whittle, G. Booch (Eds.), "UML" 2003 - The Unified Modeling Language. Proceedings, 2003. XIV, 415 pages. 2003.

Vol. 2860: D. Geist, E. Tronci (Eds.), Correct Hardware Design and Verification Methods. Proceedings, 2003. XII, 426 pages. 2003.

Vol. 2859: B. Apolloni, M. Marinaro, R. Tagliaferri (Eds.), Neural Nets. X, 376 pages. 2003.

Vol. 2857: M.A. Nascimento, E.S. de Moura, A.L. Oliveira (Eds.), String Processing and Information Retrieval. Proceedings, 2003. XI, 379 pages. 2003.

Vol. 2856: M. Smirnov (Eds.), Quality of Future Internet Services. IX, 293 pages. 2003.

Vol. 2855: R. Alur, I. Lee (Eds.), Embedded Software. Proceedings, 2003. X, 373 pages. 2003.

Vol. 2854: J. Hoffmann, Utilizing Problem Structure in Planing. XIII, 251 pages. 2003. (Subseries LNAI).

Vol. 2853: M. Jeckle, L.-J. Zhang (Eds.), Web Services - ICWS-Europe 2003. VIII, 227 pages. 2003.

Vol. 2852: F.S. de Boer, M.M. Bonsangue, S. Graf, W.-P. de Roever (Eds.), Formal Methods for Components and Objects. VIII, 509 pages. 2003.

Vol. 2851: C. Boyd, W. Mao (Eds.), Information Security. Proceedings, 2003. XI, 453 pages. 2003.

Vol. 2849: N. García, L. Salgado, J.M. Martínez (Eds.), Visual Content Processing and Representation. Proceedings, 2003. XII, 352 pages. 2003.

Vol. 2848: F.E. Fich (Eds.), Distributed Computing. Proceedings, 2003. X, 367 pages. 2003.

Vol. 2847: R.d. Lemos, T.S. Weber, J.B. Camargo Jr. (Eds.), Dependable Computing. Proceedings, 2003. XIV, 371 pages. 2003.

Vol. 2846: J. Zhou, M. Yung, Y. Han (Eds.), Applied Cryptography and Network Security. Proceedings, 2003. XI, 436 pages. 2003.

Vol. 2845: B. Christianson, B. Crispo, J.A. Malcolm, M. Roe (Eds.), Security Protocols. VIII, 243 pages. 2004.

Vol. 2844: J.A. Jorge, N. Jardim Nunes, J. Falcão e Cunha (Eds.), Interactive Systems. Design, Specification, and Verification. XIII, 429 pages. 2003.

Vol. 2843: G. Grieser, Y. Tanaka, A. Yamamoto (Eds.), Discovery Science. Proceedings, 2003. XII, 504 pages. 2003. (Subseries LNAI).

Vol. 2842: R. Gavaldá, K.P. Jantke, E. Takimoto (Eds.), Algorithmic Learning Theory. Proceedings, 2003. XI, 313 pages. 2003. (Subseries LNAI).

Vol. 2841: C. Blundo, C. Laneve (Eds.), Theoretical Computer Science. Proceedings, 2003. XI, 397 pages. 2003.

Vol. 2840: J. Dongarra, D. Laforenza, S. Orlando (Eds.), Recent Advances in Parallel Virtual Machine and Message Passing Interface. Proceedings, 2003. XVIII, 693 pages. 2003.

Vol. 2839: A. Marshall, N. Agoulmine (Eds.), Management of Multimedia Networks and Services. Proceedings, 2003. XIV, 532 pages. 2003.

Vol. 2838: N. Lavrač, D. Gamberger, L. Todorovski, H. Blockeel (Eds.), Knowledge Discovery in Databases: PKDD 2003. Proceedings, 2003. XVI, 508 pages. 2003. (Subseries LNAI).

Vol. 2837: N. Lavrač, D. Gamberger, L. Todorovski, H. Blockeel (Eds.), Machine Learning: ECML 2003. Proceedings, 2003. XVI, 504 pages. 2003. (Subseries LNAI).

Vol. 2836: S. Qing, D. Gollmann, J. Zhou (Eds.), Information and Communications Security. Proceedings, 2003. XI, 416 pages. 2003.

Vol. 2835: T. Horváth, A. Yamamoto (Eds.), Inductive Logic Programming. Proceedings, 2003. X, 401 pages. 2003. (Subseries LNAI).

Vol. 2834: X. Zhou, M. Xu, S. Jähnichen, J. Cao (Eds.), Advanced Parallel Processing Technologies. Proceedings, 2003. XIV, 679 pages. 2003.

Vol. 2833: F. Rossi (Eds.), Principles and Practice of Constraint Programming – CP 2003. Proceedings, 2003. XIX, 1005 pages. 2003.

Vol. 2832: G.D. Battista, U. Zwick (Eds.), Algorithms - ESA 2003. Proceedings, 2003. XIV, 790 pages. 2003.

Vol. 2830: F. Pfenning, Y. Smaragdakis (Eds.), Generative Programming and Component Engineering. Proceedings, 2003. IX, 397 pages. 2003.

Vol. 2828: A. Lioy, D. Mazzocchi (Eds.), Communications and Multimedia Security. Proceedings, 2003. VIII, 265 pages. 2003.